SWIMMING LESSONS

SWIMMING

LESSONS

KEEPING AFLOAT

IN THE

AGE OF TECHNOLOGY

David Ehrenfeld

OXFORD
UNIVERSITY PRESS

2002

OXFORD
UNIVERSITY PRESS

Oxford   New York

Athens   Auckland   Bangkok   Bogotá   Buenos Aires   Cape Town
Chennai   Dar es Salaam   Delhi   Florence   Hong Kong   Istanbul   Karachi
Kolkata   Kuala Lumpur   Madrid   Melbourne   Mexico City   Mumbai   Nairobi
Paris   São Paulo   Shanghai   Singapore   Taipei   Tokyo   Toronto   Warsaw

and associated companies in
Berlin   Ibadan

Copyright © 2002 by Oxford University Press, Inc.

Published by Oxford University Press, Inc.
198 Madison Avenue, New York, New York 10016

Oxford is a registered trademark of Oxford University Press

Library of Congress Cataloging-in-Publication Data
Ehrenfeld, David.
Swimming lessons : keeping afloat in the age of technology /
David Ehrenfeld.
p.   cm.
Includes bibliographical references
ISBN 0-19-514852-5
1. Human ecology.   2. Environmental degradation.
3. Civilization, Modern—21st century.
4. Technology—Social aspects.   I. Title.
GF49 E47 2001
304.2—dc21   2001032180

Epigraphs in *Swimming Lessons* have been taken from the following sources. Two selections from *Life Is a Miracle*, by Wendell Berry, Copyright © 2000 by Wendell Berry. Used by permission of Counterpoint, a member of Perseus Books Group. One selection from *The Collected Poems of Thomas Merton* by Thomas Merton, extract from "A Letter to Pablo Antonio Cuadra Concerning Giants," copyright © 1963 by The Abbey of Gethsemani, Inc. Reprinted by permission of New Directions Publishing Corp. and Laurence Pollinger Limited. Two selections from *Voltaire's Bastards: The Dictatorship of Reason in the West*, by John Ralston Saul, copyright © 1992 by John Ralston Saul. Used by permission of Vintage Books, a division of Random House, Inc. One selection from *Building a Bridge to the 18th Century: How the Past Can Improve Our Future*, by Neil Postman, copyright © 1999 by Neil Postman. Used by permission of Vintage Books, a division of Random House, Inc. One selection from *Leap*, by Terry Tempest Williams. Copyright © 2000 by Terry Tempest Williams. Used by permission of Pantheon Books, a division of Random House, Inc. One selection from *Evolution as a Religion: Strange Hopes and Stranger Fears*, by Mary Midgley, copyright © 1985 by Mary Midgley, Methuen. Used by permission of Routledge (Methuen). One selection from *China's Examination Hell*, by Ichisada Miyazaki, trans., Conrad Schirokauer, Weatherhill, copyright © 1976 by permission of Weatherhill, Inc. One selection from *The Dream of the Earth*, by Thomas Berry, copyright © 1988 by Thomas Berry, Sierra Club Books. Used by permission of Sierra Club Books.

1 3 5 7 9 8 6 4 2

Printed in the United States of America
on acid-free, recycled paper

TO WENDELL BERRY

The vertigo of the twentieth century needs no permission of yours or mine to continue. The tornado has not consulted any of us, and will not do so. This does not mean that we are helpless. It only means that our salvation lies in understanding our exact position, not in flattering ourselves that we have brought the whirlwind into being by ourselves, or that we can calm it with a wave of the hand.

Thomas Merton,
"A Letter to Pablo Antonio Cuadra
Concerning Giants"

PREFACE

When I was in medical school in the early sixties, the Cold War between the United States and the Soviet Union was at its height, and the possibility of a nuclear inferno seemed—and probably was—very real. Those were the days of the Cuban missile crisis, of constant nuclear testing by both sides, of belligerent threats and counter-threats, and of the ubiquitous presence of nuclear fallout shelter signs at subway entrances and on large, public buildings. At that time, the hands of the symbolic clock printed on the cover of *The Bulletin of the Atomic Scientists*, indicating the estimated danger of a nuclear exchange, were set at one minute to midnight. Many people were frightened; I can remember sitting with my classmate, Bill Kates, and talking about the fastest escape routes out of Boston in the event of an attack. I seriously considered moving to New Zealand, which, because it was on the opposite side of the globe from America and Russia, appeared safer to me.

As it turned out, I never gave any of my escape strategies a trial run. It's just as well; in retrospect they were pretty silly. Others did more. Edward Teller, the father of the hydrogen bomb, was alleged to have built an underground bomb shelter for his personal use. Later, according to reports, it was seriously damaged by one of the brush fires that are so common in southern California. Whether true or not, the story is plausible and makes an important point; given the large numbers of nuclear weapons in existence, their enormous power, the lasting effects of their radiation, and the likelihood that in the event of a nuclear war quite a few of the intercontinental ballistic missiles fired will go off course in unintended and unpredictable directions, there is not much that any individual, even an Edward Teller, can do to prepare to survive such a conflict. Nor is it clear that survival would be desirable.

Now, in the twenty-first century, the public fear of nuclear war has decreased. The Soviet Union is gone; there are, as I write, nuclear disarmament treaties in place; the actual number of missiles has been greatly reduced; and there is a limit to how long we can be terrified by a menace that doesn't materialize. True, there are still several thousand functioning or malfunctioning missiles left; the treaties are threatened by America's proposed development of a "limited" missile defense shield that responsible scientists know cannot possibly work; and nuclear weapons are now so compact that terrorists can smuggle them across international borders in trucks or in the kinds of small planes that are used so successfully by drug runners. The nuclear threat is still real; nevertheless, we don't seem to worry much about nuclear war anymore.

New fears have replaced the old. We are now aware of massive human overpopulation and disruption of natural habitats around the world, with resulting widespread loss of species, depletion of critical resources, pollution, and resurgence of infectious and other environment-dependent diseases. We also see the growing power of multinational corporations, a tremendous concentration of wealth, the crippling of government's ability to respond to serious situations such as global warming, the acceptance of a dependent consumer mentality, and our alienation from and ignorance of nature.

These new threats are different from nuclear war in a critical respect. They are not all-or-nothing. There is nuclear war or there isn't—no middle ground, no possibility of any sane person mistaking one state for the other. But menaces such as global warming, increasing selfishness and crass materialism, our adaptation to pathological lifestyles, the rapid erosion of local communities that hold our civilization together, decreasing biodiversity, and other modern ailments are not all-or-nothing. They are happening now everywhere around us, not so slowly that we need not worry, but not so quickly that they are manifestly obvious to everyone, including those blinded by short-term profits and the lust for power.

In a sense, most of our contemporary problems are analog, not digital in nature. A middle ground exists between "off" and "on"; this is a reason for hope. Because a little time elapses between our first awareness of these newer threats and the point at which they will irrevocably alter nature and human civilization, there is the possibility of working out protective strategies, if we can.

That is what this book is about: first recognizing the present and future dangers, and then seeing what we as individuals and communities can do about them. I don't mean bomb shelters and survival kits with water purifiers, wind-up radios, and potassium iodide tablets; nor am I thinking of plane tickets to New Zealand (or, if you already live in New Zealand, to somewhere else). I mean bringing about changes in the way we live and raise our children: changes that do not cost much money and that do not require a big infrastructure, changes that we can tailor to suit our personal needs, changes that we can enjoy making, once we get the energy and determination to try them out. In other words, swimming lessons—how to cope with a challenging environment in a healthy, gratifying, and positive way.

I can offer no guarantee of personal survival if you follow the lessons that emerge from this book. We have entered a period of exceptional upheaval, with worldwide social unrest made more dangerous by the availability of novel and fearsome technologies of destruction. Nor do the natural forces that have been stirred up by our heedless environmental disruptions—storm, drought, disease, pest—always strike in predictable, manageable ways. Yet, knowing that we may well end up in deep water far from shore, how many of us will refuse to learn to swim because we might encounter a shark?

The chapters of *Swimming Lessons* originally appeared as "Raritan Letters," the name of my column in *Orion* magazine. There has been extensive rewriting, much new material has been added, and some of the titles have been changed. I thank the publisher of *Orion*, Marion Gilliam, for providing me the opportunity to put my ideas before *Orion's* readership in the company of other writers whose work I admire, and I gratefully acknowedge permission to use this material in *Swimming Lessons*. My regular editor at *Orion*, Aina Barten, has been giving me the benefit of her remarkable editorial skill and judgment for as long as I have been associated with the magazine. I never cease to learn from her—or to be surprised at how much there is to learn. *Orion's* managing editor, Emerson Blake, has also been tremendously helpful: I value his critical reading and thoughtful suggestions. Readers of this book are encouraged to visit *Orion* magazine and The Orion Society on the web at www.OrionOnline.org, or to phone at (413) 528-4422 (Great Barington, Mass.) for more information.

In creating *Swimming Lessons,* I set myself the goal of producing a book that was more than a collection of separate pieces. This involved many changes and, in some cases, considerable additions to the original texts. Because each chapter was first written for the same editors and the same audience, this was not hard to do. I found that five themes—the titles of the main sections that follow—keep reappearing in my writing, and I used them to integrate and group the chapters in what I believe is a coherent way. This is my fourth book with Oxford University Press, and my editor at Oxford, Kirk Jensen, has done his usual stellar job of making me feel welcome and leading me through the labyrinthine process of turning a manuscript into a book. The production editor for *Swimming Lessons*, Robin Miura, and the copyeditor, Cornelia B. Wright, have been a pleasure to work with and have improved the manuscript in ways I could not have done.

I have drawn inspiration from many people. Wendell Berry, to whom this book is dedicated, has been a friend and counselor for many years; his books mark the moral center for me and many others concerned with the fate of our civilization. David Orr and Wes Jackson are other friends whose influence, both acknowledged and intangible, has helped shape *Swimming Lessons*. For approximately ten years, I have been meeting weekly with Hans Fisher, a colleague and friend at Rutgers, to discuss Jewish texts over lunch. His wisdom has, I know, found its way into these pages. Peter Raven and Naftaly Minsky corrected several errors in the text. And there are many present and former graduate and undergraduate students who have helped me with their constructive comments, information, and friendship. They include Tara Bowers, Jitendra Joshi, Meg Holden, Kristi MacDonald, Timon McPhearson, Yuko Sakano, Jonah Smith, Rochelle Smith, Jyoti Tambwekar, and Jason Tesauro.

As the readers will discover, my ideas are inseparable from my life and experiences, and this means inseparable from my children, Kate and her husband Dan, Jane, Jon, and Sam. They are woven into my stories as they are woven into my existence, both of which are enriched beyond my power to thank or describe. And as they grow older I have found their comments about what I have written increasingly valuable. Rarely can a parent have been blessed by so much wisdom and love from his daughters and sons. No doubt they take their cue from their mother, Joan, my companion and guide for more than thirty years. That Joan has read and improved every sentence of this book is the least that she has done to support me in its creation. I'm not much of a swimmer, but she is, and her presence is what keeps me above water—and even lets me enjoy the swim.

Needless to say, with editors, friends, and family like these, the only parts of *Swimming Lessons* for which I can take full credit are any errors that may have remained.

Finally, a word about the title. I took "Swimming Lessons" from the title of the last chapter. I like it because it implies, correctly I believe, that there is no need to drown in the sea of materialism and bad technology that surrounds us. Since choosing the title, however, I have discovered that there are other books out there called *Swimming Lessons*. I don't find this a problem; after all, the books themselves are totally different. I am attached to the title and mean to keep it. So to paraphrase the great British humorist P. G. Wodehouse, who ran into a similar difficulty with the title of his book *Summer Lightning,* I can only hope that some day this work may be thought worthy of inclusion in the list of the hundred best books entitled *Swimming Lessons*.

New Brunswick, New Jersey   D. E.
February 2001

CONTENTS

ONE

THE LIES WE LIVE

Never has failure been so ardently defended as though it were success.

John Ralston Saul,

Voltaire's Bastards

BRAINSTORMING

Thinking is the most overrated human activity.

Wendell Berry

A number of years ago, I participated in a brainstorming session in Washington, D.C. Although not my first panel meeting, it was, nevertheless, the first that officially had been called a brainstorming session, and I was filled with nervous anticipation as my train pulled into Union Station the night before. The next morning my anxiety increased as I sat munching Danish and drinking decaf from a disposable paper cup in the posh conference room of the environmental organization that was cosponsoring the meeting with a federal environmental agency. What was brainstorming? Would I be able to do it?

I had looked up "brainstorming" in *The Oxford Modern English Dictionary* back home in New Jersey. The first definition given was "a violent or excited outburst often as a result of a sudden mental disturbance." This didn't seem right. Nor did the second definition: "*colloq.* mental confusion." Not until I got to the fourth definition was there a glimmer of hope; it read, "A concerted intellectual treatment of a problem by discussing spontaneous ideas about it." But I was still confused, mentally confused. What was a "spontaneous idea"—or, to put it another way, what was a non-spontaneous idea? How did a discussion of spontaneous ideas differ from an ordinary serious discussion? Hesitantly, I had concluded that a brainstorming session must produce more exciting and definitive results than a mere discussion.

At first I was alone in the conference room, free to eat more than my share of Danish. Then a woman arrived. By her air of confidence and authority I knew instinctively that she was the chair of the session. We spoke, and I cleverly and casually turned the conversation to the topic of brainstorming, maneuvering to get some kind of clue about what was expected of me. But the audio technician (that's what I discovered she actually was) was more interested in the subject of endangered species and in the technicalities of the recording system. No hope. Soon the room filled with august biologists and social scientists; I sank inconspicuously into my high-backed chair, trying to breathe softly and blend into the upholstery pattern.

The session began. Our charge, as the expert panel on biodiversity, was to determine the "megatrends" that will affect the global environment in the next thirty years, and to predict which "events," "forces," "uncertainties," and "surprises" would have significant environmental impacts during that time. Right from the beginning I fell behind while musing about how one could predict surprises, and when I snapped out of it, I became aware that I was lost. The conversation didn't make sense to me—I had missed too much. Each panel member seemed to be talking about his or her specialty or personal experiences; I couldn't make out the subtle connections that were no doubt weaving the individual strands of conversation into a brilliant, unified prediction of our biospheric destiny.

Time passed. I ate more Danish and found that I was in one of the three or four working groups into which the expert panel had been divided. Now we were writing questions, trends, forces, and surprises on a huge pad of paper held on an easel. The colored markers we were using had a toxic smell: I blamed my mental confusion during this part of the brainstorming session on the fumes from those markers.

The final two hours of the meeting were devoted to reconciling the findings of the various working groups. Because of the markers, I didn't understand my own group's conclusions; soon I discovered that I couldn't follow those of the other groups either. They appeared to be statements of the obvious, and it was then I knew that the effects of the fumes had not quite worn off. We were into the moment of grand synthesis, the final spurt, and I was still floundering near the starting line. The best I could do was nod sagely and let the flashes of wit and insight pass me by.

When the brainstorming session ended, I left quietly, assuming that I had done a pretty good job of concealing my inadequacy. I was wrong.

An envelope containing the draft summary report of the meeting arrived in my office a few weeks later. I opened it eagerly. The report came to the point quickly. On the second page of text I saw four "strategic issues" printed in bold type—the first read, "The future of biodiversity will be determined by social, cultural, and historical forces." With a sinking feeling, I quickly scanned the other three bold entries. They were just like the first. The fourth stated, "A 'patchwork quilt' of constantly evolving institutions is required that can solve problems at various scales." This was followed by many pages of text and tables, including a whole page of possible "negative" and "positive" Surprises, the first of which was "meteorites." Where were the exciting and definitive results that I had expected? The report of the brainstorming session was of no practical use to anyone.

I knew it was my fault. I had failed; I had not pulled my weight. I had ruined the brainstorming session. Evidently, for brainstorming to work everyone must be having spontaneous ideas, and my ideas had hardly been spontaneous at all.

Of course, once the word about me got out it would be unlikely that I would ever be invited to another brainstorming session, but there was always a chance, and the next time I must be prepared. What was I doing wrong? I used to believe it was best to tackle environmental and other problems that I had a personal stake in, to confront them directly one at a time, and to concentrate mostly on the present where I was on firm ground, even though real progress was often slow. This way, I thought, there were time and opportunity for personal knowledge and personal emotions to inform one another, minimizing the errors brought on by pure thought or pure feeling. Here was my mistake; I had a bad attitude. I hadn't realized the power of expert planning and forecasting, the importance of thinking globally, and the incredible possibility of figuring out the earth's future after one or two days of exchanging spontaneous ideas around a conference table—given the right mix of experts and facilitators.

It was too late to fix the session that was past; however, I could improve my brainstorming skills by replaying my role at the expert workshop in the privacy of my office, trying to get it right. For nine hours, without pausing for food or drink, I practiced having spontaneous ideas about the megatrends that will affect the future of biodiversity, technology, and society. Hunger and thirst sharpened my faculties (I thought ruefully of the Danish and decaf), and after several hours, ideas began to pop into my head almost too fast to write down.

I thought about the device almost everyone will be wearing in the coming electronic age that will alert them when they are about to make the transition between virtual and ordinary reality: a device with vinyl bladders filled with dried, simulated peas, that flap the wearer gently on the ears—left ear when going into virtual reality, right ear when going out. I thought about the dawning era of global free trade and the single remaining multinational corporation, Consolidated Z, which will control all products, prices, wages, and local zoning variances. . . . The ideas were still not quite spontaneous, and there was an unsettling residue of the specific. But I was improving steadily.

Around six in the evening, a hunger spasm made me turn my head, and my eyes were distracted by a copy of an essay by the British philosopher, Mary Midgley, titled "Why Smartness Is Not Enough," lying on a corner of my desk. It was open to page 48, where I read the words, "All around us, we can see people trying to solve by logical argument, or by the acquiring of information, problems that can only be dealt with by a change of heart." For a moment I felt discouraged—had I been wasting my time in a fruitless activity? Then I stretched, took some deep, aerobic breaths, and was myself again. Midgley must have been referring to other people, not to brainstorming environmentalists. Time for one more spontaneous idea.

I thought, finally, about the last remaining wood turtle, in the spring of the year 2019, an old barren female in Potter County, Pennsylvania, tapping the moist earth with a brick-red foreleg, her wise and patient brown eyes searching for the first sign of a responding worm, while on the ridge top the yellow bulldozers gather and miles away, in Washington, the experts are still brainstorming, brainstorming.

PRETENDING

For as high as [the ostrich's] body is, yet if they thrust their head and neck once into any shrub or bush, and get it hidden, they think they are safe enough, and that no man seeth them.

Pliny the Elder, *Natural History*

My earliest memory—was I two? three?—is of a nurse or babysitter who was dressed in white and had a bad smell. When she came into my room, I would pretend to be asleep. She looked down at me (I could sense her presence), then, satisfied by my closed eyes and quiet, if shallow, breathing, would turn and leave the room. Soon her odor would go away, too, and I could breathe deeply again. This taught me a lesson of dubious value: when helpless in a situation, pretending can give you power. I, a small, weak child, had controlled the movements of an enormous, smelly adult. It was some time later, I suppose, when I learned that pretending usually doesn't work. Although I can't recall the time or place, I know that on one grim day of disillusionment and reckoning I discovered that when I closed my eyes I didn't become invisible; I couldn't transport bullies to distant, foul places by imagining them there or alter the course of unwelcome events by pretending they were otherwise.

Healthy children come to know the difference between pretending that is relaxing, stabilizing, healing, necessary—we call it fantasy—and pretending that is dangerous. A hot fire burns, even if we pretend that it won't. In our personal lives most (but not all) of us learn to instantly distinguish harmless from harmful pretending, so we do not pretend in a way that endangers our lives or physical well-being. Strangely enough, society as a whole is a different story. For at least the past fifty years, probably longer, we have been working hard as a high-tech civilization to ignore the limits of safe pretending, even to blot them out of our collective memory. And the more obvious the warning signals, the more blatant our pretending has become. A few examples will make the point.

The first example is genetic engineering. In the late 1950s, I was fortunate to have as one of my teachers a visiting professor from England, Francis Crick of Watson and Crick fame, who taught part of an upper-level biochemistry course on the structure of macromolecules, especially desoxyribonucleic acid, or DNA. Those were exciting days. Crick, a master teacher, explained his and Watson's model of DNA with brilliant clarity. The structure of DNA, the sequence of its four kinds of component molecules (bases), was like a book, in that it contained information that was then copied into a sister molecule, RNA. RNA, in turn, used this information to direct the synthesis of the many different proteins that formed or helped to synthesize all the molecular parts of the cell. This was the "central dogma" of molecular biology: DNA makes RNA makes protein. Genes are made of DNA, and the genetic "code" of the DNA molecule, it appeared, is the same for all organisms—plants, animals, microbes. Decipher the DNA code of one, several, or perhaps all the genes of a species, and the possibilities for manipulation and control of the growth, metabolism, form, and behavior of that organism are limitless. At least that's what we believed.

By the 1960s and '70s, the central dogma had given birth to the new field of genetic engineering. At the core of genetic engineering was transgenics, the semi-controlled transfer of genes from one individual to another, even across species boundaries. In the decades that followed, bacterial genes were moved into corn and soybeans; goats were engineered to express the silkmaking genes of spiders; human genes were transferred to sheep, pigs, salmon, and other creatures. Genes have been deleted, added, and modified to achieve the desired results. The very nature of a species and the boundaries between species have come to seem as malleable as soft clay. Genetic engineers now claim the prerogatives once ascribed to God. The student genetic engineering club at my college calls itself "Designer Genes."

Yet such claims fall far short of reality. Despite the enormous popular enthusiasm whipped up by the press and the financial markets, only a few of the simplest possible genetic manipulations among the many that have been tried have worked at all. These few have generally turned out to be disappointing, dangerous, or both. The central dogma concerning the importance of DNA remains largely true, and its discovery is rightly celebrated as one of the greatest feats of twentieth-century biology, but the enormous power and precise control attributed to it are mostly imaginary. It has become increasingly apparent that DNA is only a part of the story—itself subject to other regulating and modifying influences in the cell, influences that we dimly understand. Rather than being at the top of a simple linear chain of command, the DNA of the gene should be seen as one piece of an interacting complex of regulatory systems and feedback loops, with no single element "running" the cell, let alone the entire organism. One of the first of the molecular biologists to perceive this was the American geneticist Barbara McClintock, whose early work on "jumping genes," published in 1951, won a belated Nobel Prize in 1983, twenty-one years after Watson, Crick, and Wilkins received their prize. More recently, the distinguished British molecular biologist and biophysicist Mae-Wan Ho has described the biological reasons that underlie the multiple failures and dangers of genetic engineering and related technologies such as cloning.

In her very important and lucid book, *Genetic Engineering: Dream or Nightmare?* Ho explains some of the biological challenges to genetic engineering. They include the instability of transgenes, overlapping genes, gene amplification and inactivation, environmentally induced changes in gene expression, inheritance that does not involve the DNA code, and even, possibly, the inheritance of acquired characteristics. She asks, "Would anyone think of investing in genetic-engineering biotechnology if they knew how fluid and adaptable genes and genomes are? The notion of an isolatable, constant gene that can be patented as an invention for all the marvellous things it can do is the greatest reductionist myth ever perpetrated."

Ho also carefully evaluates the hazards of genetic engineering, particularly the increased risk of spreading antibiotic resistance among disease organisms and human and animal populations, and the risk that gene-transfer technologies will create new varieties of virulent pathogens. The latter is not proven, but forty new pathogenic human viruses appeared between 1988 and 1996—an unusually large number. Horizontal (between-species) gene transfer, a restricted and infrequent process under natural conditions, appears to be responsible for at least some of the new viruses. During these same years, genetic engineers were working hard to develop and apply new technologies for facilitating horizontal gene movements, using plant and animal viruses and other DNA-containing "shuttle vectors" designed to escape the usual cellular barriers against such transfer.

Commenting on biotechnology's frantic rush to create new organisms, Ho writes, "What the public is up against is a selective blindness to evidence among the genetic engineers and a single-minded commitment to look solely for the exploitable, which is the hallmark of bad science." In other words, the genetic engineers are pretending that their technology works as claimed, is stable, and is safe, that the euphoria of 1960 is still scientifically justified, in spite of what we have learned since then. Nor is the public entirely blameless in the matter. Especially in the United States, we have been all too willing to make the effortless choice of following the biotechnologists into uncharted territory, eyes and minds tight shut, delusions intact. (I would be less than fair if I failed to note that some biotechnologists are scrupulously honest about the implications of their work. For example, in an article in the March 30, 2001 issue of *Science*, M.I.T.'s Rudolf Jaenisch and the Roslin Institute's Ian Wilmut—the cloner of Dolly—describe the "drastic defects that occur during development" and the "high failure rates" that are part and parcel of the cloning of mammals, and they warn that the cloning of humans would be "dangerous and irresponsible" and will remain so for the "foreseeable future.")

Pollution control and risk management provide another example of delusional thinking. The public has been willing to pretend that the chemical industry's regular release into the environment of a huge number of biologically active chemicals is somehow being monitored and controlled in the public interest by governmental agencies. Modern management and regulation of pollution in the manufacturing and use of chemicals, however, are based on many hidden assumptions. We assume that there is a threshold concentration below which the chemical has no adverse effect or has an effect that can be considered an "acceptable risk." We assume that it is reasonable to measure and evaluate the toxic effects of each chemical by itself, independent of other chemicals. We assume that measurement of toxicity can be done by standardized tests at moderate cost. We assume that the calculated toxicities of tens of thousands of chemicals are not additive: that if each regulated chemical in the environment is at or below its "safe" concentration level, then all chemicals taken together are also safe. We assume that our regulation of toxic chemicals can be enforced—that accidents and willful violations will not make a mockery of our concentration limits. We assume that governments will have the resources, both funds and trained personnel, to test all potential environmental and health toxins in a timely and thorough way. We assume that we know of all of the environmentally important chemicals that are being released during manufacturing or during product decomposition. And we assume that our testing takes into account the different effects and fates of each compound in air, water, and soil. All of these assumptions are demonstrably wrong.

As the biologist Joe Thornton writes in *Pandora's Poison*, "The idea that science can precisely manage thousands of chemicals, calculating harmless doses and permitting exposures below these levels, radically overestimates our understanding of chemical toxicology and basic biology." In other words, much of the ponderous environmental protection system we have created to regulate the chemical industry, and nearly all of the "risk analysis" profession that has sprung up to support that regulatory process, is seriously wanting, despite the efforts of thousands of dedicated people. Cancer rates continue to rise sharply, and a flood of synthetic, hormone-mimicking chemicals continues to exert its wide-ranging effects on people and wildlife—chemicals thought to be involved in such dissimilar findings as earlier breast development in young girls, decreased attention span in some children, and problems with inner ear development and hearing in whales.

Our system to protect the environment and public health from synthetic chemicals has had its successes, but it is always playing catch-up, and will be until it adopts a fundamentally different approach. There is an alternative to what Thornton calls the "risk paradigm" presently in use. It is known as the "precautionary principle." The precautionary principle recognizes the limits of our science, and tells us, in effect, to stop pretending when the stakes are so high. In cases where there is both scientific uncertainty and a reasonable suspicion of harm, we are obliged to ban suspect chemicals until their manufacturers can prove them safe. This puts the burden of proof where it belongs, and makes pretending unnecessary.

The last example of glaring unreality is the so-called antimissile defense shield, dubbed "Star Wars" in the 1980s during the Reagan administration. Why is a substantial part of the American public willing to pretend that a limited, but fabulously expensive, shield against nuclear missiles is possible, when it manifestly is not and never will be? The computer program that will control the shield may be the most complex ever written. Complex computer programs don't run smoothly the first time they are tried, yet for obvious reasons, this one cannot be tested and debugged before it is needed. Even if, by some miracle, the control system were to function without a hitch, and every one of the interceptor missiles were to be launched at the right time, they would not hit many of the targets. This is evident from the fact that all of the important interception tests of Star Wars prototypes have failed; despite $60 billion spent on this project to date, the interceptors cannot distinguish between missiles and cheap but sophisticated decoy balloons. To hide these disappointing failures, recent tests have been rigged by the Pentagon, using poor-quality decoys that are nothing like the real missiles, and by providing the interceptor system with detailed information on "enemy" missile launch time and flight path. Even these rigged tests fail.

Theodore Postol, an arms expert at the Massachusetts Institute of Technology who worked in the Reagan administration on antimissile systems, has said that Pentagon officials "are systematically lying about the performance of a weapon system that is supposed to defend the people of the United States from nuclear attack." And Dr. Nira Schwartz, a former senior engineer at the military contractor TRW, lost her job and was investigated by Congress after challenging the company's claims that its weapons could distinguish warheads from decoys. Yet these criticisms will be irrelevant if, as seems distinctly possible, it is technically feasible and inexpensive for a rogue nation to smuggle nuclear, biological, or chemical weapons into American cities without using missiles at all.

There are dozens, hundreds of other areas of life—from the safety of our meat supply to the fairness and effectiveness of our prison system to the war on drugs to the quality and relevance of our universities—in which we posture and pretend rather than tell the truth. Why do we keep on doing this? Surely it is no coincidence that in each of these cases corporate profits would be threatened if we stopped pretending and made appropriate responses. With the massive power of the media, advertising, and political campaign contributions that they now wield, corporations encourage us and our representatives to persist in our childish delusions. We have convinced ourselves that we can totally control nature and live forever without experiencing the serious side effects of our technology; that we can use fundamentally destructive consumer goods without suffering the consequences; and that magic with a scientific face will keep us safe from our enemies. These dreams are seductive, but if we don't wake up soon, our world will wither, our security will vanish, and all our pretending will bring us no comfort.

THE MAGIC OF THE INTERNET

I am the only person in my university building who wants no part of the university's free e-mail and who does not surf the Internet, which makes me the last holdover from the days when conversations by voice, dusty books, and rectangular white envelopes with colored stamps in the upper right corner were our primary means of communication. Nevertheless, rumors of the new and wonderful electronic inventions reach me often —from my wife and children and especially from my students. Like the seven golden cities of Cibola, the Internet beckons; if I had the energy of Coronado, I would seek it out and master it. Being lazy, I sit in my office with nothing more modern than an automatic pencil, a telephone, and an aging Macintosh Centris equipped only with Microsoft Word 5.1, and think about what I am missing.

I imagine the chat rooms, for instance, abuzz with scintillating conversations, the twenty-first century equivalent of the salon of Madame de Staël or that celebrated Oxford club where J. R. R. Tolkien and C. S. Lewis read to each other from their latest manuscripts. I know that chat rooms are not really rooms, but in my mind I can see one clearly: the lustrous walnut paneling, the floor-to-ceiling mullioned windows framed by the thick pleats of burgundy velvet drapes, the comfortable armchairs upholstered in rich mahogany leather, and next to one of the chairs— my chair—a small Louis Quatorze table holding a Waterford snifter with a generous dollop of Napoleonic brandy. Sometimes the rooms are contemporary in style, with walls consisting of giant liquid-crystal surfaces that shimmer with ever-changing, abstract splashes of color or bold, functional patterns of steel-like beams intricately linked, endlessly dissolving and reconnecting. In place of chairs there are convivial arrangements of very large, "smart" cushions that rapidly mold themselves to the contours of the body regardless of whether one wishes to sit or recline.

And the conversations! Here must be the ultimate purpose for which language was invented, especially my language, English, with its unparalled wealth of words. Imagination falters as I attempt to conjure up the excitement of these electronic dialogues that I have no part in: the lilting rhythms, the novel metaphors, the memorable turns of phrase that elevate the written word above the level of mere chatter. But it's no use; I am shut out of that world by my own lack of enterprise.

This deprivation might be bearable if the *conversaziones* of chat rooms were all that I was missing, but an article in *The New York Times* of February 21, 2000, made it clear that much more was happening on the Internet than I had suspected. I don't mean auctions, news, reference services, and cheap airline tickets; I mean a solution to one of the most intractable problems faced by industrial society, the problem of energy.

The *Times* article, by one Joseph Kahn, was headlined "Surge in Oil Prices Is Raising Specter of Inflation Spike," and was subheaded "Questions over Vulnerability of 'New Style' Economy to 'Old Style' Commodity." "New Style . . . Old Style"—what were they talking about? As my eye roamed the columns of text, a boldfaced quote from the story caught my attention: THE CONVENTIONAL WISDOM SAYS OIL NO LONGER MATTERS IN THE AGE OF THE INTERNET. It took a full minute for me to grasp what I had read; the Internet had ended the country's dependence on oil. The implications were staggering. And to think all this had taken place while I, who consider myself informed, had been slumped inertly in a desk chair gazing at a tank full of guppies when I might have been on-line.

Perhaps it wasn't true. Yet I had too much respect for conventional wisdom and the *Times* to doubt it for long. Obviously it was true. The Internet had replaced oil—the only question was *how*. I had to find out.

No doubt I could have learned to surf the 'net, where the secret would have been revealed to me, but my computer is not hooked up to anything except an electric wall socket at one end and a printer at the other, and I didn't know how to get on to the Internet from either of these connections. No, I would figure out how our need for oil had been electronically eliminated by using the method I as a scientist know: by examining the world around me and asking significant questions, such as, "What uses oil? Is it saving oil because of the Internet?"

Cars, trucks, and planes were a likely place to start. For one thing, I own a car—was it using less gas lately? Not really. Nor had my online friends noticed any improvement in their mileage; in fact, one complained that he could no longer afford to fill up the tank of his giant SUV. Meanwhile, I learned that a large group of angry truckers was deliberately tying up interstate highway traffic as they slowly drove to Washington to protest the high prices of diesel fuel. If they were using less, they wouldn't have been so upset. And the airlines were scaring away customers by slapping an oil surcharge on their tickets.

It wasn't transportation; perhaps it was the use of home heating oil that had been cut by the Internet. Although a check of our heating records showed that I was still burning the same amount of oil per degree day as in past years, maybe my Internet-savvy neighbors were not. The easiest way to find out was to drive to the office of my oil delivery company, a nice, family-owned business. They were pitifully happy to see me.

"You're not screaming," the bookkeeper explained. "Everybody screams at us now. They don't realize that we aren't making big profits from these oil prices." Here too, the Internet's New Style was not in evidence.

Food production came next in my systematic survey. Again, although I kept an open mind, I found no indication that the Internet was helping agriculture use less oil to make nitrogen fertilizer, spray pesticides, run modern farm machinery, or ship apples from South Africa. So I turned hopefully to manufacturing and warfare, both thriving internationally and both traditionally heavy consumers of oil. Factories, battleships, missiles, and stealth bombers are crammed with computers, and computers are almost synonymous with the Internet. This is where the oil savings must be. The reasoning was sound, so I couldn't understand why global manufacturing and war were consuming more oil than ever.

Then, my investigation received yet another setback. In a dentist's waiting room, I read in *Forbes Magazine* that the United States is using more than 13 percent of its electric power to make and run its computers and to operate the Internet, and the percentage is growing rapidly. Of course, most of this electricity comes from coal, not oil, but it takes a great deal of oil to mine and ship the coal—oil that was hardly being used at all for that purpose twenty years ago, when computers and the Internet were in their infancy. How then could the *Times* report that oil no longer mattered in the age of the Internet, if the Internet itself was running on oil? It was baffling.

My readers, for whom the Internet has no secrets, have probably long since guessed where I went wrong in my reasoning—however, it took me many hours before I figured it out for myself. The breakthrough came in a conversation with a friend who is a stockbroker.

"It used to be," he explained, "that you could tell a lot about a stock from the price/earnings ratio. A solid company like General Motors, which makes real products and earns real profits, has a market capitalization of about ten times earnings, a p/e of 10. On the other hand, an Internet stock like Yahoo has a p/e of more than 1,500. Their stock is worth almost as much as General Motors', but they have hardly any earnings. We've had to invent a new ratio, p/s, price/sales, for those Internet companies whose stock is worth billions but which have no profits at all."

"It's odd," I replied, "that companies that make no profits should be worth so much—a kind of magical inversion of reality." And then, as I realized what I was saying, the penny dropped. I had been making the mistake of thinking in an outdated, twentieth-century mode, in which cause frequently leads to effect and things are sometimes what they seem. But this is indeed a new age of wonders; if companies can use the Internet to achieve prosperity without profits, why can't the country use the Internet to solve its energy problems? Somehow, the Internet has created an economic environment where everyone profits, nobody loses, and there is never (well, hardly ever) a final accounting. Similarly, the Internet has, in ways I cannot begin to understand, suspended the first law of thermodynamics, creating a limitless supply of oil from a limited source.

When this New Style Internet-energy will reach the gas pumps is a question whose answer I leave to the experts. I myself have no doubt that it will happen soon, and America's exciting, energy-consumptive way of life will be able to continue indefinitely. In the meantime, until gas prices get a little lower, I have started to walk rather than drive to the local supermarket. But I do this after midnight so that the neighbors won't see me—I don't want them to think I have lost faith in the Internet.

NOTHING SIMPLE

There is a saying attributed to the eighteenth-century Ukrainian philosopher Gregory Skovoroda that has long stayed in my mind. Skovoroda wrote that "we must be grateful to God that He created the world in such a way that everything simple is true and everything complicated is untrue." I have always suspected this statement to be very profound, and I would like to believe it. I often do believe it. As I write these words, I am pretty sure that I believe it. But I can't say that I am 100 percent free of doubt. Gregory Skovoroda came of simple peasant stock and lived in the days before multinational corporations, environmental management, lengthy tax returns, and global free trade. I know that great wisdom is supposed to be timeless, unaffected by changing circumstances, yet I wonder if exceptions can be made for times like ours. The trouble is that so many of the things I used to think were simple have turned out to be complicated. At times, usually around six in the morning when I am still in bed and life seems grim, I even wonder whether anything is simple anymore.

Take the case of overpopulation. Any ecologist worthy of the name knows two things about overpopulation: first, it is bad; and second, the United States, because of its excessive per capita consumption, is already overpopulated. In this country, ecological, social, political, cultural, and economic systems have all been strained past the breaking point by the burden of too many people consuming too much.

True, most citizens do not yet perceive the magnitude of the danger. The root causes of our worst problems rarely appear in plain view. As the biologist and social critic Garrett Hardin once wrote, "Nobody ever dies of overpopulation"; the death certificate—if there is a death certificate—lists something more tangible such as tuberculosis or AIDS. It is difficult to acknowledge distant ultimate causes such as overpopulation when there is something easier and closer at hand to blame. Nevertheless, increasing numbers of people, even some economists, are beginning to recognize the simple idea that the world and its wealth are finite and that population growth must stop.

But are things really so simple? I don't know—there is a complication.

The perception and judgment of overpopulation seem to be culture-dependent. National comparisons do not work when population is concerned. Take China and the United States, for example. The United States, with more than 280 million people, is feeling the stress of crowding—from the national parks to the expanding suburbs. So, not surprisingly, is China with around 1.3 billion. But what was China like in the early decades of this century when it had one-fifth of its present population, 250 million? There is less habitable land in China than in the United States; was Chinese overpopulation in 1920 similar to that in the U.S. in 2001? I doubt it.

The Chinese, then as now, ate little meat and had, compared with us, few large farm animals and none of the massive waste of calories and protein that occurs when grain is fed to cattle, so their efficiency of food production was much higher than ours. Nor did the Chinese ever use as much energy per capita in keeping themselves clothed, housed, and amused. And China, as befits a country that has been densely populated for a very long time, has over many centuries evolved customs and codes of social behavior that make it easy for large numbers of people to rub elbows without society unraveling in chaos, arson, and bloodshed.

Prominent among the cultural traits that have enabled large populations to live together in China has been a long-standing sense of self-discipline combined with an awareness of the narrowness of the social pyramid as it tapers toward its apex. This applied to the privileged as well as the lower classes. Until the early years of the twentieth century, a Chinese boy of a cultivated family was taught his first twenty-five written characters shortly after his third birthday. They spelled out an encouraging but cautionary verse (translated by Conrad Schirokauer, from *China's Examination Hell*, by Ichisada Miyazaki):

> Let us present our work to father.
> Confucius himself
>   taught three thousand.
> Seventy were capable gentlemen.
> You, young scholars,
>   eight or nine!
> Work well to attain virtue,
>   and you will understand propriety.

By the time he was fifteen and ready to take the imperial examinations for a government position, the boy had memorized texts containing 431,286 characters. (Meanwhile, his sister was learning a different and even more demanding kind of discipline and submissiveness.) These imperial exams, which were instituted in the sixth century, continued to be given until 1905. Echoes of the system persist in contemporary, densely populated Japan and in other parts of crowded Asia. In the United States, by comparison, fewer social adaptations to crowding seem to exist. As we push on toward a population of 300 million people, we still retain a fixed idea of personal space so grandiose that it was wasteful and destructive even in frontier days, when there were fewer than a third as many Americans as there are now. This is the kind of intolerance of crowding that led citizens of New Jersey to murder each other for a better place in the gas station lines during the "energy crisis" of 1973, and which has reappeared in other forms since then.

In addition to evolving social adaptations to their dense populations, the Chinese have developed ecological habits that reconcile them, as far as it is possible, to the environment in which they are crammed. Although anything resembling wilderness vanished from the inhabited parts of China ages ago, the Chinese have at least come to terms with their farmland (until recent modernization), as evidenced by four millennia of farming without loss of soil fertility, and the invention, more than two thousand years ago, of integrated pest management, using cultivated citrus ants to keep pests out of orchards. The United States, on the other hand, has relied primarily on a heavy-handed technology to provide enough food for a growing population (and for export), with disastrous environmental and social effects too numerous and too familiar to be listed.

If, as seems likely, the definition of overpopulation is complicated by a national relativism, with each country or culture experiencing a different threshold, how can we say that this is a simple idea, and therefore true? Would Gregory Skovoroda, having followed the argument up to now, turn his back and walk away in disgust? Or could we persuade him that the basic idea of overpopulation is still a simple one? "Gregory," we would say to him, if American, or "Mr. Skovoroda," if Chinese, "wait a minute. Don't be in such a rush to leave. Overpopulation is not a complex notion even if there are a few qualifications. After all, what could be simpler than the statement that we need one-child families for the world to survive? One child, Gregory (or Mr. Skovoroda)—that's what the world needs. You can't get any simpler than that." Or can you? Leaving the Ukrainian philosopher to ponder our words by himself, let's step out of earshot and look more closely at the simple idea of the one-child family.

When I was growing up I had a younger sister, but a four-year difference in age, plus sibling rivalry, put her beyond the pale for companionship as far as I was concerned. Fortunately, for I was socially backward and had few friends, I had seven cousins in the neighborhood, three of whom I enjoyed playing with. My favorite was my cousin Elaine, or Laney, who was almost my age and who in addition to being lively and interesting lived around the block from me. Our backyards touched at one corner, making visits convenient; although to get from one yard to the other in the easiest way we had to cross through the flower beds belonging to an elderly couple who yelled very loudly.

Laney and I taught each other to play table tennis using a kind of elongated card table with a makeshift net and, in lieu of real paddles, the hard backs of books that we had salvaged from the garbage. I have never had as much fun using proper equipment. Whether Laney, now an august professor of statistics at an East Coast university, still plays table tennis, I don't know. I also remember the parties, mostly at cousin Bobby's house, where Bobby, Jimmy, Laney, and occasionally Billy, who was older, all vainly tried to help me conquer my fear of girls.

My childhood was enriched by cousins because my father happened to have four married siblings living nearby. My cousins on my mother's side lived in faraway places such as Minneapolis and Miami—I rarely saw them. But my parents and all of their surviving siblings, a total of nine brothers and sisters, were extremely close regardless of the physical distance that separated them, sharing happiness and supporting each other in every way during the inevitable times of trouble. This sense of family was the dominant influence during my early years.

In traditional China, family was of yet greater importance. Not only were ancestors worshipped and elders respected, but relatives of every age and degree were recognized and significant in each person's life. As Charles O. Hucker wrote in his history, *China's Imperial Past*, "The individual was never alone in his confrontations with the world at large; he was representative of, responsible to, and supported by his family." The Chinese language itself has words for various kinds of family relationship that take whole phrases to describe in English. Even in modern China, this dominant fact of the family as the center of social existence has persisted in spite of the one-child policy begun under Mao.

And what of the one-child family? In a nation of one-child families, what words drop out of daily language? "Brother" and "sister" to begin with—there are no brothers and sisters in one-child families. Because there are no brothers and sisters, there can be no "aunts" and "uncles," and therefore no "cousins," neither first nor second nor once or twice removed. No "sisters-in-law," no "brothers-in-law." Blood may be thicker than water, but in a one-child world, water will have to do.

Of course many people grow up happy and well without siblings, cousins, aunts, or uncles. And some extended families are stifling and terrible. It is perfectly possible to have good, close relationships with nonrelatives during childhood and afterward; nearly everyone does. But an extended family can offer a richness—and sometimes a stability—that is hard to duplicate in modern societies, whether affluent or impoverished. In affluent societies such as ours, most social forces seem to move the individual toward an ever greater isolation that is masked but not compensated by the riotous consumption of material goods. Families can be the best antidote to isolation. In poor societies, an integrated family is often the only barrier between an individual and political oppression or death from starvation.

Thus the passing of the large family, necessary as it may be, is not the simple demographic blessing that the population planners would have us believe. What will happen to us without it? If we are lucky, social evolution under the pressure of necessity will replace the old family with a new kind of sustainable social structure. Perhaps it will be one in which single adults, childless couples of assorted sexual combinations, and parents with various numbers of children somehow share a formal "family" structure that offers support to all. Many utopian books set forth such schemes, and some exist, but I doubt whether social structures that are deliberately and consciously contrived—engineered—can survive for long in most places, or become widespread. Complex social systems need some evolutionary trial and error, which is a lengthy process, to make them stable and sustainable in the real world.

In the meantime, the human population is growing and growing. The earth groans under it. There is no place to hide. There is nothing simple. If Gregory Skovoroda is right, everything that we believe about our society and our future is untrue because it is so complicated. Or perhaps I have misapplied or misunderstood his words . . . perhaps he wasn't referring to things like overpopulation and other modern problems . . . perhaps some meaning was lost in translation from the Ukrainian. . . .

SHERLOCK, NERO, AND US

Some people have a drink when they get home from work—a martini, a beer. Maybe two or three. Life is especially stressful in the twenty-first century. The same indecent forces that are destroying nature are disrupting our working lives as well. Who will own the company tomorrow? Will there be a "reduction in force" or some other euphemism for the ax? When is the next reorganization coming? Is my ten or twenty years of faithful service an asset or, more likely, a sign of obsolescence and suspicious loyalty to bosses and coworkers now out of favor or working for other companies? How am I to answer the fax that arrived at 4:00 p.m., the one that seemed to contradict the fax that arrived at 11:00? When will I find time to fill out the questionnaire from the Resource Management Office, and does it take precedence over the Goals Enhancement Strategy questionnaire that came from the Administrative Services Division? Are my computer files compatible with the new software system, and, if so, why did the box on the screen say, "This paragraph is unreadable. Do you wish to substitute a standard paragraph?" Just what is our real work, anyway?

Alcohol can take the edge off stress, but it is not everyone's consolation. Some choose television or the Internet; mine is to pick up one of the scores of detective novels that I keep close to hand and plunge in. Then I can forget for a little while the vice presidents, deans, and other academic, corporate-style bosses who do their best to make life in the modern university an unproductive misery. In this way I can put out of mind, temporarily, the pleas of students who don't understand why there are no courses to take and the ravings of colleagues who can't figure out how to cope with the contradictory, impossible demands placed on them.

Why should detective stories be, for so many, such a good and entertaining way of escaping from reality? That they are is clear; billions of copies have been sold. What's more, the fictional detectives endure decade after decade, outliving their creators, even their creators' children. In the real world as time passes, gas lamps give way to carbon arc lights, which are replaced by tungsten bulbs, which are supplanted by mercury vapor lights, halogen bulbs, and a host of fluorescents; yet Sherlock Holmes still glides through the lamplit streets of Victorian London as the wisps of yellow, pea-soup fog gather and drift in countless bedrooms on six continents.

Modern imitations of Conan Doyle's Holmes and Rex Stout's Nero Wolfe—some actually bearing the names of these immortals—claim valuable shelf space in today's bookstores, but copies of the original stories outnumber them. The character of the great Chinese detective, Judge Dee, embellished and reinvented by many centuries of Chinese mystery-storytellers, and in our time brilliantly resurrected and recreated by Robert van Gulik, has survived since Dee's actual death during the Tang Dynasty, at the end of the seventh century. Hercule Poirot still twirls his waxed mustache and sips sickeningly sweet sirops, outselling the competition in the days of purple hair and margaritas. And the Hardy Boys and Nancy Drew, though they are now no strangers to jet planes, remain most compelling and most popular in books written when gearless bicycles and ancient Fords were the only transportation available to these adolescent sleuths.

We may feel that our stress is greater than that of past times; indeed, it is only during the last forty to fifty years that we have developed the most destructive technology and the most devastating overpopulation in human history. But the detective story, with many of the same detectives who comforted our grandparents during their troubles, is still capable of comforting us. This calls for an explanation. I have one.

What the detectives of fiction offer to us is confirmation by example of an idea that is becoming ever more in conflict with the reality of our lives and ever more precious: the idea that there is an order and sense to the happenings around us, and that this order can be discovered if we know how. We have the possibility of understanding and controlling events.

Television and the Internet give us similar illusions, for those who can tolerate them.

The detectives' secret, at least for the best of them, is simple, elegant, and—once you know it—elementary. To understand the inner workings of our complex, seemingly chaotic surroundings, the great detectives move outside of society, retaining only a narrow, carefully defined and managed connection with the world of unfathomable events and confused people.

Sherlock Holmes established the pattern. When we and Dr. Watson first meet him, in *A Study in Scarlet*, he appears at once to be an odd sort: "A little queer in his ideas" and "desultory and eccentric" in his studies, he has nevertheless "amassed a lot of out-of-the-way knowledge that would astonish his professors." As Watson's roommate, Holmes proves fascinating and occasionally congenial, as long as the infinitely obliging Watson respects his privacy and the arrangements he has made to keep the world at arm's length. Watson has amazing tolerance for Holmes's moods, both manic and depressive, for his use of cocaine, for his secrecy, for his impatient arrogance, and for his apparent lack of ordinary human emotions. This last point needs some qualification. In a handful of instances, perhaps most obviously in "The Adventure of the Three Garridebs," when Watson receives a flesh wound from a bullet, Holmes does fleetingly betray an affection for Watson.

The lure for Watson, apart from that curiously strong kind of friendship that can occasionally spring up between opposites, is the same lure that attracts us. Holmes finds sense and pattern where we see only a meaningless jumble. From a battered hat found in the street, Holmes deduces that the unknown owner is an intellectual gone to seed during the past three years, middle-aged, sedentary, out of training, whose wife has ceased to love him, etc., etc. With the passage of more than a century since "The Adventure of the Blue Carbuncle," many of Holmes's deductions seem improbable or preposterous to our sophisticated eyes. But, like Watson, we are still happy and flattered to spend time with him.

With Nero Wolfe and his assistant, Archie Goodwin, Rex Stout took the Holmes concept and, in the four decades following 1934, revised it for the age of atomic bombs, television, administrators, and consumers. Stout's achievement, a monumental one, was to show that it might be possible for a person to be both successful and free in our society; he demonstrated—if only in fiction—the power that can come from skillful and uncompromising avoidance of the entanglements and commitments of commercial culture. To do this, Stout moved Wolfe farther from society than any detective before him.

The big difference between Nero Wolfe and us is that Wolfe sets his own daily schedule and orders his own life. Self-employed, naturally he determines his hours, but I don't mean that he works half-days or takes Tuesdays and Thursdays off. His day begins around eight, when his cook, Fritz Brenner, brings breakfast to his room. From nine to eleven in the morning and from four to six in the afternoon, Wolfe is in the greenhouses on the roof of his Manhattan brownstone, supervising his gardener, Theodore Horstmann, in the care and breeding of a vast collection of choice orchids raised for his amusement and pleasure. If clients try to see Wolfe while he is in the plant rooms—no matter how urgent their business or how influential they are—Archie Goodwin sends them away or makes them wait. Likewise for district attorneys and police inspectors. Murder and mayhem may swirl around his doorstep, but at the times he has appointed for himself, Wolfe remains serenely ensconced among his orchids.

Nor is any business discussed at meals, which are long and lavish (the Nero Wolfe stories have spawned at least one cookbook). Wolfe refuses to leave his house on business and seldom leaves for other reasons; he prefers the society of his household and few close friends, and, moreover, distrusts and loathes cars and planes. He rarely shakes hands, never makes small talk, and takes on cases only when it suits him and he needs the money. He does not rely on gadgets and technology to solve crimes; his knowledge of what is going on in the world comes primarily from newspapers, not television.

Wolfe is massive (one-seventh of a ton, we are told), formidable, and even less given to expressing emotion than Holmes. There is no way to relate to him except at his discretion, on his terms, and then across a barrier. The only people he respects are those who, as he does, define themselves without regard to outside influences. Nero Wolfe dictates his own needs—the ultimate antimodern man. In Wolfe, Rex Stout has created that rarest of types in our time, the true conservative, one who insists that there are unchanging standards by which all things, from food to progress, must be judged.

The fatness of Wolfe—also a rejection of our society—is as principled and deliberate as the other parts of his existence. In his house, the preparation and eating of food is a cultural statement, religious in intensity and serving some of the same separating and identifying purposes that food serves in traditional Jewish, Hindu, and Islamic life. Once, in the book *In the Best Families*, Wolfe is forced to flee his house and change his identity. He becomes part of the culture he spurns, subsists on the typical American food he detests, and loses a huge amount of weight. His acquired thinness is more than a disguise—it is a way of impressing the reader with the magnitude of his sacrifice and the connectedness of the different parts of a principled existence. Once the threat is removed and he can go home, Wolfe's first act is to call the farmer who supplies him with poultry and instruct him to resume feeding select chickens on blueberries. Only then do we know that things are back to normal on West Thirty-fifth Street.

It is Archie Goodwin, Wolfe's legman, who connects Wolfe with the world around him (and with us), making the detecting plausible and the books so much fun to read. Archie revels in the culture and amenities that Wolfe rejects—the fast cars, the fashionable, expensive suits, and the baseball games. To say nothing of the dancing at the Flamingo Club. (The outspoken misogyny of Holmes and Wolfe is another way of setting them apart from the world and of making things simpler and therefore more fathomable. It is consistent with their characters, but it is not a necessary ingredient in detective fiction.) Archie enables Wolfe to manipulate and relate to a society that Wolfe chooses not to enter. Wolfe needs Archie, but what allows Wolfe to tolerate him is Archie's integrity of character. Like Wolfe, Archie is self-defined; he sets and keeps his own standards. The two men understand what it takes to live uncorrupted in the modern age, so they understand and appreciate each other.

Both Nero Wolfe and Sherlock Holmes are answerable first to themselves. If their principles coincide with the law, as they usually do, they uphold it. If not, they get around it. In "The Adventure of Charles Augustus Milverton," Holmes and Watson break into a private home, where they deliberately destroy evidence that would identify a murderer whom they believe should not be punished. In *The Next Witness*, Wolfe evades a subpoena and an arrest warrant to find a criminal his own way. There are many other instances from the accounts of these detectives and others. A notable example is *Mr. Pottermack's Oversight*, by R. Austin Freeman, in which the scientific detective, Dr. Thorndyke, solves a murder that has baffled the police; he then decides not to tell them who did it. In all such cases, the detectives share the perception that fidelity to people, places, and principles can be more important than the formal requirements of law, and that the law is not a license to suspend personal judgment and responsibility.

Both Conan Doyle and Rex Stout, like the celebrated detectives they invented, were originals, people of exceptional integrity and talent who had proven to their own satisfaction that it was possible, with skill and care, to change the world around them. Doyle was, besides a writer of fiction, an ophthalmologist, successful inventor, dramatist, war correspondent, whaler, athlete (he introduced skiing to Switzerland), historian, and spiritualist. Stout, a child prodigy in arithmetic, was a naval officer (on President Teddy Roosevelt's yacht), an itinerant bookkeeper, sightseeing guide, stable hand, hotel manager, inventor of a widely used school banking system, radio commentator, prominent opponent of Nazism and nuclear weapons, cook, and (like his sister, Ruth Stout) a brilliant pragmatic gardener. Nobody could describe them as narrow or unworldly. That they each chose to distance their detectives from the worlds of their time can only be decisions based on wide experience.

In Sherlock Holmes and, especially, Nero Wolfe, I have picked extremes to make a point. Most other fictional detectives move more comfortably through the world about them. But nearly all have some distinguishing characteristic that lets them step away from and make sense of their environments. Here are a few examples of the best: Sergeant Jim Chee (Tony Hillerman), a religiously committed Navajo, is trained in anthropology; Mr. Campion (Margery Allingham), Lord Peter Wimsey (Dorothy Sayers), and Lovejoy (Jonathan Gash) live on the fringes of their respective societies; Jane Austen (Stephanie Barron), a detective based on the early nineteenth-centruy novelist, is a woman in a man's world; Inspector Ghote (H. R. F. Keating) and Superintendent Otani (James Melville) have extraordinary standards of personal honesty which set them apart; Inspector van der Valk (Nicolas Freeling), unimpeachably Dutch, has a perceptive and embarrassingly unconventional French wife; Brother Cadfael (Ellis Peters) is a Benedictine monk with vows of renunciation to observe. Each of them discovers and restores the underlying pattern in a world that makes sense in spite of its ambiguities, in spite of how chaotic and frustrating it seems to us ordinary folks.

But the world, especially our world of the twentieth and twenty-first centuries, does not make elementary sense, and it makes less and less with every passing year. The idea that it can be figured out and re-arranged by exceptionally talented individuals is a comforting lie. There are a few modern fictional detectives who know this, who admit and ex-plore the unpredictability, complexity, and uncontrollability of our world, and who in the end must fail. These detectives, like Adam Dal-gliesh (P. D. James), suspect that there may be no "great underlying pat-tern." The considerable enjoyment their adventures provide is that of the good novel, not the reassurance of the classical mystery. The differ-ence is real, as Conan Doyle and Freeling discovered when, having killed off their detectives, they were forced by their readers to bring them back again.

It is pleasing after a day at the job to step silently and invisibly into Nero Wolfe's office, stand next to the great globe, and dream that there really are people like Wolfe who can figure out what is going on. The in-satiable demands of global management in our increasingly self-con-structed world cry out for a Wolfe, a Holmes, a Poirot—people who can discern the patterns and gather up the myriad, tangled strings that make the marionettes dance. But if we are honest, stop lying, and look around us at the cascading problems caused by our prior solutions, we can see that no human, and certainly no machine, is up to the task of fixing them single-handedly. We never were, of course, but before the age of science most were willing to admit our ultimate shortcomings.

Could there be even a modern *fictional* equivalent of Wolfe in this dis-integrating culture? Yes, I think so, but with changes. What will the next Nero Wolfe be like? For one thing, he will surely disdain e-mail and the Internet. For another, he will probably be a she. I'm waiting eagerly. The twenty-first century takes a lot of escaping.

TWO

WRECKING OUR CIVILIZATION:

A MANUAL

Is this the curse of modernity, to live in a world without judgment, without perspective, no context for understanding or distinguishing what is real and what is imagined, what is manipulated and what is by chance beautiful, what is shadow and what is flesh?

Terry Tempest Williams, *Leap*

REJECTING GIFTS

To forget our mistakes is bad. But to forget our successes may
be worse.

Neil Postman,

Building a Bridge to the

18th Century

A long-remembered scene repeated many times: I am about sixteen; it is
5:45 P.M. and we are at the dinner table, with my mother at one end, my
father at the other, and my twelve-year-old sister, Judy, sitting across
from me. As always, the meal is formal—the oval mahogany table is covered with a clean tablecloth and there are three courses: appetizer, main
course, dessert. And as always we are eating very quickly to keep up with
my father, who must be back at the office at six to see more patients. The
phone rings; I can hear our maid, Josephine, going to answer it. A
pause. Then Josephine comes into the dining room. "It's Dr. Williams,"
she tells my father.

My mother is annoyed. "Why does he always call at dinnertime?" she
asks, but my father is already on the way to the phone. Dr. Williams and
my father, both practicing physicians, occasionally meet at Passaic General Hospital, but Dr. Williams, who has heart problems, is also my father's patient, and that is why he is calling. He knows that my father puts
his patients ahead of his dinner. Soon, my father returns to snatch a
hasty bite of dessert and tells my mother, "I'll be going to Rutherford

after office hours tonight. I should be back around ten." My mother says nothing.

Another scene, equally well remembered: My father and I are in a car, and he is driving; I am not yet old enough for a license. He speaks to me in that tentative, half-apprehensive tone of voice that fathers use toward adolescent sons when they are about to propose something new. "You know that I have a patient, Dr. Williams, over in Rutherford? Well he's beginning to go blind and needs somebody to read to him once a week for an hour or two. I told him I would ask you—he would pay you for your time."

From my hesitation, my father knows that I am not liking the idea. "You would enjoy reading to Dr. Williams," he continues. "He's a very interesting man. He writes poetry." But it's no use. I make my excuses: too busy, too much schoolwork, too inconvenient. The fact is, I just don't want to read to an old friend and patient of my father. A few days later the request is repeated, then a week after that. By this time it has become a matter of principle and I am adamant. Eventually my father stops asking.

A third scene, just a few days before this writing: Dr. Williams is long dead; my father is dead, too. I am sitting in my living room holding a familiar book from my father's collection. A slender volume, elegantly printed on creamy, rough-cut paper yellowing slightly at the edges, it is the fourth book of a set, which together make up one of the great epic poems of twentieth-century English literature. Opening it carefully to avoid tearing the fragile dust cover, I read that this is the first edition of Volume IV of *Paterson*, of which one thousand copies were printed for the publisher, New Directions, by Van Vechten Press, Inc., Metuchen, New Jersey. And I see once more the inscription by the author, written in blue ink: "Irving Ehrenfeld, gratefully yours, William Carlos Williams, 9/19/51."

Despite the painful regrets I have, I don't blame my sixteen-year-old self for rejecting the marvelous gift offered to me by my father and his celebrated patient. Now, with two adolescent sons of my own and two daughters who were adolescent not so long ago, I understand that becoming independent sometimes involves the rejection of gifts. It is a necessary cost of human development, a tragic cost, because so many of the gifts can never be offered again. But it is a cost that must be borne— there is no alternative.

The theme is one that I think about often because we live in a time when the rejection of gifts is commonplace, a true way of life. Our cultural landscape is littered with fragments of gifts rejected unnecessarily, and it is the adults, not the adolescents, who are doing the rejecting. The techno-economic system that dominates the industrial world has little patience for the bearers of evolved wisdom, for anyone encumbered by connections with the past and concerns for the future.

This was brought home to me recently when I met Dr. Oved Shifriss on the grounds of Rutgers's Horticulture Farm No. 3; I hadn't seen him in about a year. Retired from the faculty for more than a decade, he must be in his eighties. The shapeless, faded blue canvas hat was the same, but he is now wearing hearing aids in both ears. The senior farm worker who was with him, much younger and obviously very busy, waited patiently and respectfully while we spoke—or rather while Dr. Shifriss spoke and I listened. Oved Shifriss, son of Israeli pioneers, once the chief of research for Burpee Seeds and probably the greatest vegetable breeder Rutgers has ever had, knows as much about the genetics of squashes, cucumbers, melons, and pumpkins as any person on earth. When I first came to the university, I was surprised to notice that even his long-time colleague, himself an accomplished scientist, always addressed him as "Professor Shifriss." I soon understood why.

Shortly before his retirement, Dr. Shifriss told me the reason for his decision to leave. He had recently won a major award in the All-America vegetable trials for his Jersey Golden Acorn squash, but he wasn't satisfied. "This is only the beginning," he said. "If they gave me a team to help me—a few geneticists, one or two plant pathologists, an entomologist, a biochemist—I could produce a squash that would feed the world. It would be about the size of a sweet potato, very productive, insect-resistant, maybe with silvery leaves, highly nutritious, rich in Vitamin A, protein, and calories, and very delicious. I can do it, if the administration would only give me some resources and leave me alone.

"But they don't seem to be able to grasp the idea. You know how they criticized me when I was working with the little ornamental gourds. 'Those gourds are bitter and inedible,' they said. 'Why are you wasting the Experiment Station's money on them?' But I got the B-gene complex that enabled me to breed the Jersey Golden Acorn partly from the same gourds. It tripled the amount of carotene. And look at the trouble they give me because I am doing some studies with castor beans—the plant that Jonah sat under in the biblical story. 'They're poisonous,' they keep telling me, 'what good are they? You're supposed to be a vegetable breeder.' But castor beans have an extremely unusual sexual system; it's going to be important one day."

By the time he retired, not long after that conversation, Dr. Shifriss was spending a sizable part of his own salary to replace the research support that the Experiment Station had withdrawn. The super squash that he envisioned and could have created has never been developed; there is probably nobody left with his combination of knowledge, experience, and genius to do the job. Genetic engineers from Europe have shown interest in his castor bean studies. In fact, castor beans are why I ran into him at the farm: he is still allowed to keep a couple of his special plants in one of the greenhouses there. We talked a bit more—he was explaining some of the reasons for the high failure rate of genetic engineering in agriculture, and his ideas about the unexpected variability he has seen within cloned plants that are supposed to be identical—then, tiring, he said good-bye and walked slowly off toward the greenhouse.

As I watched him go, I thought of the terrible waste of his talents and wisdom. Our demand for immediate gratification and quick, carefree returns on all investments leaves no time to wait for intricate things to be worked out, no time to evaluate true worth, no willingness to listen to those who contend with natural complexity. Knowledge earned during long years of study of people and nature is being cast aside to make way for technologies and financial systems that must satisfy their own, not human, needs.

Another example: a few years before Oved Shifriss left Rutgers, Professor Fred Hough, one of the world's great apple breeders, told me about the trouble he was having with the administration of the Experiment Station. "They want me to start submitting progress reports every six months! I try to explain that I'm breeding *trees*, not radishes—things don't change in a six-month time frame. But they don't get it."

Dr. Hough, too, had a grand plan. "I can breed an apple tree with good-tasting apples that will be largely resistant to the common pests and diseases. It won't require all these chemical sprays. I can put my hands on every bit of the genetic material I need. It will take a research group of nine people—that's not much for a university of this size. But all they care about is growing a bigger, shinier Red Delicious." Dr. Hough retired in disgust without embarking on his great project. When I last heard of him years ago, he was consulting on fruit tree breeding for Romania and Brazil.

Not only do we reject the gifts of genius present, we also spurn the offerings of genius past. Gene Logsdon, farmer and author, has written extensively about the brilliant methods our agricultural forebears worked out for using the gifts provided by nature, what modern ecologists would call "ecosystem services." A choice example is keeping chickens and cows togther in the same farmyard. The chickens pick the faceflies and horseflies off of the trusting and cooperative cattle, giving the cows relief while gaining nourishing insect protein in their own diet. A different kind of service is provided by the dung beetle, which feeds her larvae on little balls of manure she buries in the ground. In two days, a pair of dung beetles can bury an entire cow pie, which if left on the surface would smother the grass and attract cattle parasites. Underground, the manure serves to nourish the soil and to produce more of the useful dung beetles. The service is free; the cost of rejecting it is not.

These gifts have been rendered irrelevant on most modern farms. Chickens are raised in crowded factory facilities—with no freedom to move, with beaks cut off so that they can't peck each other, with artificial illumination night and day—and they are fed a semiliquid diet synthesized from grain products and additives. Meanwhile the cows are also packed into enclosures and feedlots; their pests and parasites are controlled by insecticidal sprays and chemicals added to their diets. The same chemicals, passed through in the manure, are probably responsible for the widespread decline of the dung beetles, which pretty much disappeared around 1950. And the manure, itself, is now a pollutant, not a nutrient.

My stories of rejected gifts have been taken from agriculture, but I could have found them in any walk of contemporary life, from architecture to engineering, from business to medicine to education. The life of contemporary civilization is like the fairy tale of the three little pigs and the big bad wolf run in reverse. Having started with a house of brick, we have moved, with great fanfare, to a house of straw; it is hard to concoct a happy ending for such a story.

We have brought this misfortune upon ourselves; there is no wolf, no malign outside force to blame. The earth is full of gifts, the perennial gifts of nature and the gifts of human creativity worked out over thousands of years of trial, error, and sacrifice. In our insistence that we must reinvent and manage the whole world from scratch, and do it quickly, in our confidence that money and power will replace, with something better, anything that once worked well, we have cut ourselves off from the bounty available to us. But the money is dwindling and the power is waning. Soon, those who remember the gifts we have rejected will try to recover them. Will they be in time? Or, like my chance of reading to Dr. Williams, will the gifts be gone beyond recall?

ADAPTATION

adapt   v. 2. intr. & refl. [to] become adjusted to new conditions.

The Oxford Modern English Dictionary

When my wife Joan and I were newly married, we lived in a north Jersey suburb not far from the New York state line. Every weekday morning we drove down the Palisades Interstate Parkway to the George Washington Bridge and crossed the Hudson River to Manhattan, where I taught and Joan was a graduate student. The parkway runs along the Palisades, a magnificent, igneous bluff that flanks the west bank of the Hudson and faces, on the far shore, Yonkers, the Riverdale section of the Bronx, and Manhattan. Wooded parkland extends on both sides of the road for its entire length until just before the approach to the bridge, where many lanes of superhighway converge on the toll booths. We loved the woods along the parkway—they calmed us before our immersion in the chaotic city, and soothed us when we left it at the end of the day.

That was before we went on our honeymoon, a three-week hike on the Appalachian Trail (interspersed with some hitchhiking on country roads), from Springer Mountain, Georgia, to the border of the Great Smokies in North Carolina. The forest we walked through was a mixture of tall pines and an incredible variety of native hardwoods—an experience of natural diversity that was overwhelming. Nearly every tree we saw was new to us, yet we could feel the pattern and cohesiveness of the forest as a whole. Rhododendrons formed a closed canopy over our heads, fragmenting the June sunshine into a softly shifting mosaic of dappled patches. We stepped on a carpet of rhododendron petals.

The trip was over all too quickly. The plane carrying us back descended through a dense inversion layer of black smog before touching down on the runway at Newark. Home. We were depressed and silent. The ride from Newark Airport to our house took us on the Palisades Parkway. For the first time, we became aware that the woods along the parkway were dominated by thin, ungainly *Ailanthus*, with their coarse (and, we knew, rank-smelling) foliage, and by other weedy species such as the lanky *Paulownia*. Suddenly, these exotic species seemed very much out of place. The second-growth forest looked raw, monotonous, and badly assembled. We grew more depressed. It took about a month before the Appalachian memory faded and we became accustomed once again to the Palisades parkland. Then the *Ailanthus* disappeared into the leafy background, the flowers of the *Paulownia* gave a welcome splash of purple against the green, and the disrupted state of the forest strip was all but invisible to us after the days in Manhattan.

I have described a very benign and useful example of adaptation. Although I drive the Palisades Parkway infrequently now, I'm glad it is there and grateful to those who created it and maintain it. But what about our other adaptations, the majority of them, the thousands of small and large adjustments that we make to the changes that technology and crowding are imposing on us and our surroundings?

My teacher, the great naturalist Archie Carr, once remarked (I think that he was quoting someone, perhaps Aldo Leopold), "Someday we will all learn to love gas stations." In a similar vein, while musing about the question "What will be pleasant in [the year] 3000?" he wrote in 1964 in the preface to his book *Ulendo*:

Our race is frighteningly malleable. *Adaptable* we would rather call it. Already masses of people have been conditioned to processed cheese, to bread that I think is made of plastic foam, and to gaudy fruits from which all taste has been jettisoned. . . . People can be made not just to live shoulder-to-shoulder in tiers, but to enjoy living that way.

Orwell said pretty much the same thing in the final pages of *The Road to Wigan Pier*, first published in 1937. He attributed the worst adaptations in modern life to "the decay of taste" caused by our acceptance of and demand for more machine-made articles, and by our consequent distancing of ourselves from nature. "Mechanization," said Orwell, whatever its advantages," is out of control." If we extend Orwell's thoughts, taste—that outdated notion of an absolute standard of quality—may well have been our last cultural barrier against indiscriminate adaptation.

Adaptation is usually a healthy and necessary process, as the apostles of progress frequently observe, but passive, heedless adaptation is dangerous. Surely some adaptation is needed if we are to preserve our sanity in this rapidly changing chaos of our own making. People who live in large cities, for instance, have to make many behavioral adaptations, such as suppressing their natural eye contact with strangers, if they want to survive. There comes a point, however, when the hidden cost of adaptation gets too high. I can give examples, some of which I will expand on in later chapters.

We are adapting to enforced helplessness. We used to be able to fix things ourselves; now, working parts are often sealed and can only be thrown away when broken, or require specialized tools to repair. At the same time, there has been a marked decrease in the durability of many products. We have adapted to this rapid, planned obsolescence with hardly a whimper.

We are adapting to less personal contact with others. Synthetic voice answering systems, optical scanners in high-speed checkout lines, automatic tellers, self-service gas pumps, "distance learning," computer-assisted work done in the isolation of home, and e-commerce all take the flesh-and-blood people out of our daily lives. We are getting used to doing without them.

We have already become adapted to the television screen as the source of much of our news, entertainment, and companionship. In other words, we have replaced two-way communications involving our full participation with one-way reception that allows no effective response. At the same time, we have become adapted to the steady substitution of commercials for content in the programs we watch. Our tendency to adapt to a commercial environment is often cynically exploited, as is the case with National Public Radio, whose management is slowly increasing the number and frequency of commercials on air time that was once promoted as public-supported and commercial-free.

We are adapting to less courtesy and consideration from others. This is partly because we are adapting to the imposed pressure of time on all our actions. Efficiency has become the only criterion of excellence. Courtesy takes time without yielding immediate benefits. Similarly, we are adapting to the constant demand for quick decisions. No longer is there time for reflection when we are faced with complex issues. Communities cannot survive without courtesy and time for reflection.

We are adapting to a relentless erosion of our privacy, the contraction of the boundaries of personal space in our business and social relations. The cell phone not only has intruded on the privacy of others in restaurants, meetings, and even the theater, but the users themselves have adapted to the fact that everyone within hearing distance will learn about their personal affairs.

We have become adapted to an indoor climate regulated so that it is the same during winter and summer.

We are adapting to the headlines that tell us that business executives are making many millions of dollars a year even as their companies lay off tens of thousands of workers.

We have adapted nicely to the conventional modern wisdom that there is no special reason for people to live near the place where they were raised and where the bulk of their family resides.

And we are adapting to the removal of nature from our lives and its replacement with symbolic substitutes. The nature special on television replaces the walk in the woods; the voyeurism of ecotourism replaces the joy and toil of the garden. The change of seasons, the migration of swallows, the shower of the Perseids, and the lamentable decline of the dung beetles go by unnoticed.

Although our easy adaptibility is not a primary cause of our troubles, it is allowing damaging forces to do their worst with little hindrance. Our passivity to the manipulation of our culture by short-term, antisocial, economic interests is making us pushovers for destructive change. What is more serious is that our children are growing up to accept our adaptations as baseline, natural behaviors. They have internalized them.

What can we do to stop ourselves from adapting to conditions that should not be tolerated? There is a Human Rights Watch that reports on rights abuses around the world. Maybe we need an Adaptation Watch to warn us of dangerous lapses of taste and to remind us of truly self-interested, community-fostering behaviors that we are losing. We will have to speak out loudly, publicly, and often to point out the lethal changes in our environment and rouse the sleeping memories of what we were and the awareness of what we are fast becoming.

In the end, however, the war against dangerous adaptation is a personal struggle. Each of us must set his or her own limits of the tolerable—which changes to our world we will accept and which ones we will resist. I drive a car; I don't watch television. I occasionally send a fax; I am not part of the university's e-mail system. These choices are not arbitrary, although they may be disputed. What matters is that they are *my* choices.

Becoming the arbiters of our own adaptations often means selecting a less popular and more inconvenient way of doing things. Enormous forces backed by vast resources push at us to adapt. But in resisting, we are not without power. We have the power of knowing that we can fight successfully without having to win the whole war, and the power of knowing that nature and community are on our side.

FORECAST: CHILLY, OVERCAST, LIGHT DRIZZLE, NO PEOPLE LEFT

I used to enjoy listening to the National Weather Service forecasts on my short-wave weather radio. An endlessly repeated taped message updated every few hours might be less than thrilling, but the voices of the half-dozen or so forecasters made it come alive. Each one had an identifiable style and intonation; it was easy to assign personalities, even faces, to them.

Ten years ago the announcers were all men. There was the one I labeled the grand elder, with his pontifical voice and distinctive, rolling rhythms. When cost-cutting forced the station to move from Manhattan to the grounds of the Brookhaven National Laboratory, way out on Long Island, he disappeared from the airways. Perhaps the daily commute on the Long Island Expressway was too much for the old fellow. I am sure I wasn't the only listener to mourn the loss of his avuncular cadences.

Another announcer who appealed to me spoke fluently until he came to an American Indian place name such as Manasquan or Wanaque (both in New Jersey). Then he hesitated. I could imagine the look of terror in his eyes when he scanned the next line of the script, and there it was, a word with fearsome Q-sounds or daunting combinations of consonants and vowels. If I had had any way of getting in touch with him, I would have comforted him by explaining how lucky he was to be broadcasting in the New York–New Jersey metropolitan area. Up in northern Maine, the forecasters have to cope with names such as Caucomgomoc and Chemquasabamticook.

Some announcers proclaimed their individuality with what seemed like deliberately odd pronunciations of common words. The most original was the fellow who figured out a new way to say "climate," an achievement I would have thought was impossible. He did it by lengthening the separation between the two syllables and heavily stressing the second: "cly-*matt*."

Eventually, the Weather Service hired its first woman announcer, a welcome addition; she made her mark immediately by shortening the phrase "Here are the latest Central Park observations" to "Here is the latest Central Park." This bothered the linguistic purist in me, but when all the other announcers took up the abbreviated phrase, I saw the wisdom of her choice; by simply eliminating one word and changing another, she had instantly broken through the glass ceiling and become a member of the club.

The only announcer who could keep me listening attentively through several cycles of the tape came on the air some time in 1998. She was an obviously very young woman with—a first for the weather station—an unmistakable Hispanic accent. In the beginning, I was irritated because she spoke quickly and I couldn't catch all the words. But there was a quality of sweet and wistful earnestness about her voice that made me suspend judgment: I decided to give her a chance. Over the next few weeks I noticed that she was slowing down and was becoming perfectly understandable. Maybe I'm getting used to her accent, I thought. But I later learned that the weather service broadcast staff, in response to a flood of angry letters (some viciously racist), had been working hard with this young Puerto Rican intern, coaching her on pace and pronunciation. Soon I found myself tuning into the station even when I knew what the weather was going to be, just for the sheer pleasure of listening to her.

Who would have dreamed that a weather forecast could be so charming, could evoke images of lost youth and vanished innocence, could whisper promise of a better world to come? Her voice did that for me, and when she pronounced the name of my state "New Yersey," I could hear it through her ears, against a background of surf on a Caribbean beach, with a gentle breeze rustling the fronds of the swaying palms. Of all the announcers, she was the one I liked most.

Then on Tuesday evening, February 16, 1999, everything changed. I switched on the weather radio and there was a new voice, a man's, deep, also accented, and strange. The rhythm of the speech was Scandinavian, I believed, but when I recalled the voices of Swedes and Norwegians I had known, they seemed different from this one. Every phrase and sentence was a distinct unit, with a pause of exactly the same length wherever there must have been a comma in the script, and a slightly longer pause after every period. Within phrases, the words were often run together, and some words were slurred, as if the announcer were drunk. Words came in predictable bursts, utterly devoid of emotional content. I could not imagine the announcer's face, but his voice was grim. It was also intensely boring. I heard the tape twice without taking in what he said. I never did find out what the weather was going to be.

The next day I tuned in expecting one of the morning announcers, but there he was again, deadly, inhuman. Suddenly the penny dropped. The voice sounded inhuman because it *wasn't* human — I was listening to a computer! I felt as if I had received a blow. The day was ruined for me, regardless of the weather.

Why was I so upset by a fake voice on the weather station? That evening, my friends Mischa and Eugenia explained it to me. Mischa recalled the words of the Abkhazian writer, Fasil Iskander, who observed that in Abkhazia, where the farming and pastoral population is always outdoors and where the weather is a daily matter of life and death, nobody listens to the forecast on the radio. But in Moscow, with its sheltered apartments, office buildings, and public transportation, where the weather means much less in the lives of people, everybody listens. Listening to the weather report affords us an opportunity for nostalgia, says Iskander, a connection with a world of nature that most of us no longer experience in a vital way. There is no way that a computer-generated voice can inspire that kind of nostalgia.

A few days later, I spoke with someone at the National Weather Service, in Upton, New York. Yes, he said, it was a synthesized voice. No, nobody had been fired—yet. But yes, the point was eventually to save money through downsizing; that's what Jack Kelly, the Assistant Administrator in charge of the NWS in Washington, wanted. Yes, the voice was awful, but it would be improved. They were working feverishly on the software. Well, yes, there had been complaints. They expected them to die down in a month or two. That has been their experience with computer-generated voices elsewhere in the country: after a month, people stop complaining, they adapt.

I don't want the complaints to die down, and I don't want the voice to be improved. Even if they could get the thing to sound like Laurence Olivier or Emma Thompson, I would still want it scrapped. Voice quality is not the issue, being human is, and that can't be synthesized either by computer science or biotechnology.

Most of us live a lifestyle that largely cuts us off from the weather. Now, we are not even allowed the nostalgic pleasure of listening to a live person—a person whose dreams and fears animate her voice—tell us about the sunshine or rain expected tomorrow. The mechanical announcer is an abomination, and I am angry at Mr. Kelly and the other administrators of the National Weather Service for imposing it on us.

Two thoughts console me, however. I can still go outdoors now and then to see and feel what the weather is like. And those bloodless listeners who aren't bothered by a synthetic announcer don't really need to hear about the weather anyway. So take care, Mr. Kelly. In the downsized, lonely, and unnatural world you're helping to synthesize, there may be no place for you, either.

PSEUDOCOMMUNITIES

To begin with a little anecdote. There is a classroom, an ugly, badly shaped, windowless room in a modern university building designed with students not in mind. In this room there is a small class, my class. We have rearranged tables and chairs in a semicircle around my place to defy the terrible ambience and to allow all twenty-five students to see and hear each other and me. Class is in session; I am talking. Two students sitting together in the front row—a thirty-year-old man with a pager on his belt and a twenty-year-old woman—are speaking to each other and laughing quietly; they see that I am looking at them and they continue to laugh, not furtively or offensively but openly and engagingly, as if I weren't there. I don't know what they are laughing about. Both of these students will eventually receive an A in the course for their exceptionally fine work.

As I speak to the class, a part of my mind is thinking about these students and wondering why they are laughing. Is there chalk on my face? Is my fly open? Have I repeated myself or unconsciously misused a word or inverted a phrase? Then I remember something Bruce Wilshire wrote, and the paranoia fades. The laughter has nothing to do with me. Bruce, a philosopher who works in a nearby building on campus, described the same attitude in his own classes in his book, *The Moral Collapse of the University*. This passive and casual rudeness, a fairly new phenomenon, has a simple cause, he said: "I sometimes see students looking at me as if they thought I could not see them, as if I were just somebody on their screen." When my students were laughing, it didn't occur to them that I would be bothered—at that moment they were treating me as if I were a face on television.

That's what it is; I am sure of it. The students (at least most of them) and I inhabit different worlds, even when we sit in the same room. In my world, the people I speak with are real; if I offend them they are hurt and angry *with me*, if I give them pleasure they smile *at me*, if I bore them they find an excuse to move away *from me*. We are alive to each other.

The world of my students is far more complex, a hybrid world, a world in transition. For them, some of the old world survives but it is confusingly intermingled with the artificial world of electronic communication. Up to now, this has been mostly one-way communication; the recipients are physically and mentally passive. The faces on the television screen speak and speak—my students have been watching and listening in some cases for five hours a day or more, since the age of two or three. I cannot compete with this. Three hours of class a week for fourteen weeks is no time at all compared with the television they have seen. Their authority figures are two-dimensional and they cannot hear. They neither take offense nor do they rebuke; their brief utterances are well suited to the wandering, superficial mentality fostered by the ever-flickering tube.

The world of television, by inducing passivity and unresponsiveness, has cut many of the human threads and connections that once bound people together into working communities. Lewis Mumford compared life in front of the television screen to life in a space capsule, frightening in its absolute isolation. Passivity and alienation are not communal virtues.

To the extent that television has weakened American communal life, it has weakened communal power, the only effective power to limit consumption, pollution, and the degradation of nature. This would appear to lock us into a hopeless spiral of decline, for television promotes these very evils while weakening the communal ability to resist them.

If this were all the threat that electronic communication has to offer, we might conceivably find ways to cope with the challenge, formidable as it is. We might, for example, take advantage of the fact that television, which has helped to ruin communities, has not eliminated the desire of most people to be part of a community. A lonely consumer, however passive, is a dissatisfied consumer—unstable, even rebellious. Could we use this instability to fashion a revolt against television? Not any more. Not in the age of e-mail and the Internet.

Perhaps unconsciously, the developers of electronic communication have come up with the perfect counter-strategy to prevent us from using the loneliness of people caught in the television culture to wean them away from the tube and back into the community. This counter-strategy is the creation of *pseudocommunities*, my word for assemblages of electronically linked people. Pseudocommunities have arisen as substitutes for the real ones that are going or gone. Pseudocommunities are making

everyone feel good again, are replacing enervating passivity with a sem-
blance of activity, creativity, and choice; but they are keeping the reality of true neighborly, communal responsibility and judgment far away.

Among the most familiar, albeit relatively primitive, systems of electronic, two-way communication, precursors of the pseudocommunity,
are the recorded or electronically voice-simulated phone operators. Few
people are left in the United States and Canada who have not had their
blood pressure raised five or ten points by a patronizing recorded voice
saying, "If you want to discuss your bill, press one now; if you want to
speak to a nurse to schedule an appointment, press two now; if this is an
emergency, press three now; if . . . ." Making fun of this is like shooting
a sitting duck; I leave it to those who earn a living as nightclub comedi-
ans or humor columnists. But a few words will not be amiss. These systems are a threat to communities, paving the way for pseudocommunities, in that they accustom us to dealing with facsimiles of people in our
daily lives. Whether people or facsimiles do a better job is irrelevant—
there is more to life than maximizing the efficiency of daily transactions
(although electronic operators rarely do that). Daily transactions between real people are one of the things that can make life worth living.

There is usually nothing that one can do about recorded and simulated voices except hang up, which is not always practical. The ruse of
pretending that I don't have a touch-tone phone is already being challenged by the demand that I speak the number of my response to the
electronic voice decoder. Only in the case of the artificial information
operator at 411 is there any remaining possibility of satisfaction. When
the bright, phony voice says, "What city please?" I answer "wuszch
wuszch wuszch" in a low monotone. This baffles the computer and a
real operator picks up immediately. How the phone company will deal
with this I don't know, but they will find a way. Perhaps a sublethal zap
of electricity through the receiver's earpiece will modify my Luddite
behavior.

Lately, when the wonders of the age of information and communication have got me down, I have revived my spirits by rereading three exceptional books, Jane Austen's *Emma* and Elizabeth Gaskell's *Wives and
Daughters* and *North and South*. These books take place almost entirely
within a few miles of the houses of the central characters, and they describe the incredible subtlety and wealth of interactions, for good and

evil, that one experiences in a real community. In the sort of communities described by Austen and Gaskell, the passions and activities of love, hatred, sexuality, compassion, selfishness and intellectual intercourse, modulated by and expressed through the life of the community, take on a complexity and richness that is not found in an electronically facilitated pseudocommunity. In a later chapter, I will give Jane Austen some of the attention she deserves, but here I must get back to pseudocommunities.

It seems that almost every advance of our technology brings more social disintegration. Consider interactive video communication, which is already a part of distance learning and which may some day be a part of many telephones. To the voice in the receiver has been added the face on the screen, and this changes everything. The feature that has kept the conventional telephone from destroying communities is the *lack* of visual information that accompanies the voice. The disembodied voice is a constant reminder of what the telephone really brings about: communication between people who are actually, demonstrably, perhaps distressingly distant from one another. Like a letter, a phone call is received in private from someone who is elsewhere. Add a picture and the privacy and sense of distance are disturbed, replaced by an illusion of proximity, a mockery of context. This is another step on the road to pseudocommunity.

But the danger to communities of interactive video is trivial compared with that of e-mail and the Internet. Here I fly in the face of conventional opinion; this technology is supposed to be a liberating force in society. Using electronic communications, one can send a message to any person in the world who is in the network, or direct it to large groups of people simultaneously. What better way to promote the free democratic exchange of ideas, to create a "global community"?

The grave problem with e-communication—a different problem from the well-known phenomenon of abuse and addiction—is that it fosters the sensation of being part of a community of people working, creating, and playing together for the common good. But the sensation is only that, for at the end of the day when you in Vermont and your electronic correspondent in western Texas go to sleep, your climates will still be different, your landscapes will still be different, your soils will still be different, your local environmental problems will still be different, and, what is most important, your neighbors will still be different, and while you have been creating the global community you will have been neglecting them.

The speed and simplicity of communicating electronically can be alluring and habit-forming. Unlike ordinary letter writing, anything that comes to mind can be conveyed instantly with little bother, and can receive an instantaneous response. Tens of thousands of streams of consciouness are flowing through the cables. But this is not an advantage; it is a problem. In a proper, durable relationship, many thoughts, after careful reflection, should be left unsaid. Careful reflection takes time and sometimes privacy, assets that we have stupidly wished away. Equally important, if we hope to make valid judgments about things and people, we must have information from all the senses, information that can never be conveyed fully by words, or even pictures, on a monitor. The disastrous failure of many electronically formed love affairs, once the couple finally meet face to face, is a case in point. There is no easy, glamorous way to be part of a community. The phrase "global community" is an oxymoron. Pseudocommunities detract from the real work of community-building, which, although deeply gratifying, requires a painstaking, persistent effort and perpetual learning that last as long as one keeps on breathing.

Communication is good and necessary, but not at the expense of communal integrity, which requires a balancing measure of isolation of one community from another, and, at times, of one individual from another. Electronic communications systems lack this balance, this subtle regulation of communal function. In the pseudocommunity of e-mail and the Internet it is becoming harder and harder to maintain the kind of personal boundaries that add strength and diversity to real communities and keep them from flying apart.

Soon there will be new products and systems of electronic communications on the market. It does not take much imagination to envision the various combinations of interactive video, mobile miniature cameras, holographic display software, and e-mail that are likely to be devised. Some already exist as prototypes at Bell Labs and elsewhere. Reality is being replaced with virtual reality. Where will this end? For it will end in the not-too-far-off future. It will end because the global pseudocommunity is and will increasingly become economically unstable. For the great majority of users, electronic communications do not help to create real, durable wealth. The human and natural resources devoted to these electronic systems do not result in any socially acceptable outpouring of *necessary* material goods and services. Instead, production, local economic stability, and communal security are sacrificed to transient efficiencies, destabilizing luxuries, and the quick profits of distant entrepreneurs.

Although they do not create real wealth, the new communications enable it to be shifted rapidly from place to place. Under these circumstances of global trade (especially global free trade), instantaneous global finance, instantaneous global exchange or theft of wealth-producing ideas, and facilitated global exploitation of distant resources, it is becoming impossible for most real communities—and people—to continue the slow accumulation or even maintenance of assets that is a condition of survival. Once our real communities are mostly bankrupt, the accumulated wealth and social order that support the information research, the costly electronic hardware and software, and the whole network will collapse.

The new systems of communication will also be rejected for social reasons. In addition to being exploitative and expensive, consumers rather than generators of wealth, pseudocommunities are also thin and, above all, unsatisfying. Although it is fun to play bridge simultaneously with five hundred partners in ten countries, mere fun does not sustain any but the shallowest of existences. Moreover, the loss of real human contact, combined with the breakdown of defined boundaries of self and community, will not, I think, be tolerated by most of us forever. Already, research is showing that those who spend the most time talking on computer are among the loneliest people in our society.

In an article in *The New York Times* about campus e-mail (November 11, 1996), reporter Trip Gabriel wrote that on many campuses, electronic communication is preventing the development of meaningful, communal relationships.

> Dormitory lounges are being carved up for clusters of computers, student unions are declining as gathering places, and computer-wired dorm rooms are becoming, in some cases, high-tech caves. . . . James Banning, an environmental psychologist at Colorado State University who surveyed some 100 university housing officers last year, remarked: "Universities are saying: 'Oh, my God, they're

in their rooms. How can we ever build a sense of community . . .
if they don't come out?'"

In one extreme case, a student described by Gabriel communicated
with his roommates by e-mail even though they were sitting a few feet
apart in the same room. But another student, who had become dissatisfied with electronic socializing, said, "It's easier to just meet someone. You learn how much of a difference it makes to see someone in person and actually talk to them."

I remember a photograph of a baby monkey in a psychological experiment; it was being raised in isolation with a surrogate "mother"
made of wire covered with terrycloth. The monkey was clinging to the
device but it looked profoundly sad and anxious. Electronic forms of
communications are still new and exciting. Indeed, for some who are
infirm and handicapped, electronic communication can provide a life-
saving source of human contact. But for most of us, in the end, as the
dust gathers and the glamor fades, our pseudocommunities of silicon and
plastic and liquid crystal will prove no more comforting and no more
nurturing than a surrogate mother of wire and terrycloth.

OBSOLESCENCE

At the end of the Cretaceous period, the last dinosaurs disappeared from
the earth, setting off an evolutionary jubilee among the Milquetoast-like
mammals that survived them, and preparing the ground for what was to
become, 65 million years later, a permanent source of gainful occupation
for scientists whose job it is to wonder why the dinosaurs died out. Scores
of reasons have been given for this remarkable concatenation of extinctions. Global climate and sea level were changed by a city-sized asteroid
striking the earth near what is now the Yucatan, or by a massive set of
volcanic eruptions, or by the solar system passing through the core of a
giant molecular cloud, perhaps colliding with a supercomet loosened
from the Oort cluster, which orbits the Sun beyond Pluto. Theories of
catastrophic extinction abound. Some of the most daring even conjure

up the specter of an unseen companion star to our Sun, named Nemesis, whose eccentric orbit brings a wave of potentially deadly comet showers—and extinctions—every 26 million years. But there are also paleontologists who argue that the dinosaurs went away gradually, not suddenly, over a period of millions of years, and that toward the end they coexisted with the earliest hooved mammals, including ancestors of horses, cows, and sheep. If extinction was gradual, a different line of thought opens up: perhaps the dinosaurs died out because they couldn't adapt and compete in a changing world. The big lummoxes were obsolete.

I heard about the dinosaurs' obsolescence back in my student days. It was as satisfying a notion then as it is today, especially if you didn't think about it too hard. Here were these lumbering, pea-brained reptiles, barely able to walk and chew gum at the same time, while all around and underneath them, cleverly hiding behind clumps of primitive vegetation and cleverly burrowing in tunnels in the ground, were the nerdy but smart little mammals about to emerge from the shadows and begin their ascent to glory—somewhat, it occurs to me now, like Bill Gates in the waning days of heavy manufacturing.

Unfortunately, if one clutters up this elegant picture with a few more facts, the outlines begin to get blurry. Some of the dinosaurs were small and presumably agile, possibly warm-blooded, and maybe even feathered. They had much better eyes than the mammals had then or have today, if we can judge by the dinosaurs' surviving relatives, the modern reptiles and the birds. (This is not to say that the mammals didn't *start* with good eyes, inherited from their own lines of reptilian ancestors. But they lost them during endless eons of subterranean and nocturnal skulking, and when they came out into the sunshine again they re-evolved, almost from scratch, a jerry-built substitute.) As for cleverness, if brain size is a key to evolutionary survival, what happened to the Neanderthals, whose brains were bigger than ours? Why did they become extinct? Finally, although it seems almost churlish to point it out, consider the alligators and the crocodiles. We could call them obsolete: they are big, they have small brains, they are slow and ungainly on land, and they are distant relatives of dinosaurs. Yet with a little protection from over-hunting, they can do surprisingly well in our unforgiving world.

Actually, I haven't the faintest idea why there are no more dinosaurs, and if you are reading this to find out, you can stop now. The questions that interest me more are, where did the idea of obsolescence come from, and why has it become so popular? Does its insistent presence in our lives help or hurt us? Is it possible that our constant fear of becoming obsolete is leading us along the very road that the dinosaurs traveled?

Some sense of obsolescence is, I suppose, an inevitable by-product of developing technologies. Foot soldiers in biblical times learned about obsolescence when they first did battle against men in iron chariots. George Sturt, working in his wheelwright's shop in Victorian England, knew all about obsolescence, the "immensity of changes" in machine technology that instantly "separated employers from employed . . . robb[ing] the latter of the sustaining delights which material used to afford to them." But seldom if ever in history has obsolescence become such a leitmotiv in our lives as it has in this electronic age. Anything that is gone, even the dinosaurs, is now described as obsolete.

Obsolescence today implies much more than the simple recognition of what is new. It demands the absolute rejection of the old, as if the new were only validated and confirmed by the denigration of everything that came before it. To call some object, process, idea or person obsolete is to render it beneath contempt, incapable of improvement. What force can have shaped such a destructive attitude, which at a single stroke cuts us off from our past and devalues our soon-to-become-obsolete present? One force powerful enough comes to mind: corporate greed given wings by advertising. As General Motors' Alfred Sloan and his successors have long known, obsolescence sells products. The fear of becoming obsolete is being carefully taught at this moment to countless consumers around the world, drummed home by all the energy of the media and reinforced by every branch of our education industry, from kindergarten to graduate school. "Let them eat the future," Wendell Berry once wrote. How many of us are left, I wonder, who share with Berry the precious knowledge that the future is inedible?

Selling the creed of obsolescence is hard work because obsolescence is an unnatural concept, especially when applied on a time scale of only months, years, or even centuries. By unnatural, I mean that it does not really describe what happens in nature. Consider the American chestnuts, just one hundred years ago the dominant tree of the northeastern hardwood forests, which are now teetering on the edge of extinction.

They were brought low in a decade by an introduced fungal parasite—did that make them obsolete? Extinction is often the result of a rare and highly specific event in the life of a species; the passing of that species is not a general indictment of its quality. The idea of obsolescence, as currently employed, is too vague and value-laden to be of use in biology, and it certainly doesn't apply to a diverse group of animals such as the dinosaurs.

Would we have been so quick to describe the dinosaurs of the late Cretaceous era as outdated if we did not live in such an obsolescence-conscious age? The dinosaurs no longer exist, hence the dinosaurs were obsolete; they were losers. What a strange thought, considering that more than 99 percent of all the species that have ever lived are now extinct. It makes about as much sense as saying that baroque music is obsolete, somehow inferior to the classical style that evolved from it, which itself became obsolete at the dawn of the romantic period.

Does it do any harm to apply the unnatural idea of obsolescence to nature? Probably. I am not worried about the dinosaurs. They are gone. But what about the frogs, the rhinoceroses, and the less conspicuous species of endangered nature? How tempting to write them off as obsolete (especially rhinos)—losers unable to go with the flow. Not only does this attitude undermine the conservation of vanishing species, but it distorts our perception of our own place in nature. We pretend that we are no longer part of the grand scheme, but are sitting on the sidelines judging the participants. It is a fine show for a little while, yet in reality we have been given neither permission nor power to remove ourselves from the parade. We will be judged along with the rest, by criteria that are certain to be more complex than anything we can understand or predict, and nothing like our simplistic notions of winning, losing, and becoming obsolete.

Think about *Limulus*, the horseshoe crab, that curious marine invertebrate whose overfished remnants are even now, as I write, clambering awkwardly out of the sea along the Jersey beaches, taking advantage of the spring tides to leave their eggs in the sand in the darkness of a new moon. If we knew them only as fossils in the Burgess Shale, we might say, "What preposterous creatures. How inefficiently they must have moved, with those ponderous shells and thin legs, dragging that grotesque spike. They were born obsolete, an evolutionary dead end. No wonder they are gone." But the horseshoe crab is still here, looking much as it did 50 and 100 million years ago, still bearing a strong family resemblance to its eurypterid ancestors who have not swum the seas these 400 million years.

Horseshoe crabs and alligators teach us that the word *obsolete* comes much too easily to our tongues. Even within the limited context of our own technology, we should be careful where we paste that label. Did it work? Did it last or did it create the conditions for its own regeneration? Was it beautiful? Did it give pleasure? Was it a critical link in a larger process? Did it *need* to be replaced? Is its replacement an improvement? Is it possible that it will be needed for our survival in the days ahead? Slap! On goes the label— *obsolete* —and none of these questions matter, none are asked. To call something obsolete boasts an omniscience we do not possess, a reckless disregard for the deep currents of history and biology, and a supremely dangerous refusal to look at the lasting scars our technology is gashing across our planet and our souls.

SOCIAL EVOLUTION VERSUS
SUDDEN CHANGE

The threat or promise of change brings out the frail nature of mankind's psyche. And sudden change is an imposition of instability. The rational argument, from its modern beginnings, has tried to avoid dealing with this reality. The multitude of abstract social models—mathematical, scientific, mechanical, and market based—are all based on an optimistic assumption that a schematic reorganization of society will be good for the human race. . . . [H]uman beings do not respond effectively to this sort of manipulation.

John Ralston Saul,

Voltaire's Bastards

In an address to the Pontifical Academy of Sciences, meeting at the Vatican on October 22, 1996, Pope John Paul II accepted the theory of evolution, thus bringing to an official end, for the Catholic Church, the most bitter and most persistent of all debates between science and religion. "New knowledge," John Paul said, "has lead to the recognition of the theory of evolution as more than a hypothesis." He qualified his statement somewhat by pointing out that there are many readings of evolution, "materialist, reductionist, and [his preference] spiritualist interpretations." Still, we must not quibble; the Pope has endorsed slow evolutionary change, Darwinian evolution, as the likely way that nature modifies all living creatures, including all human beings. Fashioning us in the image of God, the Pope appears to believe, took a very, very long time.

The dust has not yet settled on the great evolution war, nor will it settle soon. A few intelligent scientists are still not convinced that evolutionary theory explains the species richness of our planet and the amazing adaptations, such as eyes, wings, and social behavior, of its inhabitants. There also remain powerful religious orthodoxies that show no sign of giving up the fight for creationist theology.

A second war about evolution is now being waged, an invisible, unpublicized struggle between a different set of protagonists; it is a war whose outcome will affect our lives and civilization more directly than the original controversy ever did. The new protagonists are not science and traditional religion; instead, they are the corporate apostles of the religion of progress versus those surviving groups and individuals committed to slow social evolution as a way of life.

To understand this other struggle, it is necessary to look at evolution in a broad context that transcends biology. Not just a way of explaining how the camel got her hump or how the elephant got his trunk, the idea of evolution can also be applied to the writing of a play that "evolves" in the mind of the playwright or the "evolution" of treaties, banking systems, and anything else that changes over time in a nonrandom direction. More to the point, it applies—in nonbiological ways that Darwin inspired but probably never dreamed of—to the evolution of relationships among people in organizations or communities, one of the most important evolutionary forces affecting our lives, and the most poorly understood.

In the past, the concept of evolution has often been misapplied for political purposes. Social Darwinism—roughly explained as the belief that the rich and powerful deserve their success because they are more fit than the poor and weak—is the most glaring, dangerous, and scientifically foolish example. Nevertheless, there is a legitimate application of evolution to social processes, namely, the evolution of institutional and communal relationships, which is clearly analogous to biological evolution. Formerly, in a business, university, or community that remained undisturbed for a number of years, connections and relationships among organized groups and individuals formed slowly, the accumulated result of countless small decisions and chance events, similar to the mutations and genetic recombinations at the heart of much biological change. If one of the connections or relationships was grossly dysfunctional, it was usually eliminated by fiat or by common consent, in the same way that natural selection acts on an individual organism. And like biological evolution, which has resulted in a great many false starts, stagnant lines, and extinctions, the evolutionary pace and direction of evolving institutions and communities have often been inadequate to keep them from extinction, too. Evolution is not a kindly process, but overall it works; some species, some institutions, some communities always manage to survive and prosper if evolution is allowed to run its course. Without social evolution, it seems likely that no human systems could work at all.

In our time, strong and ruthless antievolutionary forces are at play. Social evolution is not being permitted to run its normal course. These forces, however different they may be, have a common effect: frequent, sudden disruption of established relationships, customs, and behaviors among people who work and live together. Cataclysmic changes—ice ages, abrupt reversals of the earth's magnetic field, huge volcanic eruptions, and the impacts of giant meteorites—have disrupted biological evolution in former ages of the earth, wiping out vast numbers of species in mass extinctions. One or two of these, in terms of the number of species lost, make the disappearance of the dinosaurs seem trivial by comparison. Now, human activities on a global scale are causing another mass extinction, and in a terrible extinction of a different sort, the evolving social order of our relationships, institutions, and communities is being disrupted by the powerful, inhuman forces of modern technology and economics. Changes occur too rapidly, too extensively, and too often to permit social evolution to work properly; increasingly it breaks down altogether. This process is more fundamental and dangerous than mere random destruction because it undercuts the ability of the entire society and its institutions to develop the enduring human cooperation necessary for survival. Evolution requires a critical blend of change and stability: change to provide new possibilities for coping with an ever-altered environment, stability to ensure that successful adaptations, whether new or old, are not immediately swept away and discarded in a continuous, blind, senseless upheaval.

Reorganization is the most violent of the antievolutionary forces now loosed upon the industrial world and much of the Third World as well. Once an infrequent event, a last resort to be called upon for a system that had patently failed, reorganization has become both a routine management tool and an incidental consequence of the mergers and takeovers that dominate today's business. As Scott Adams, cartoonist of the antievolutionary nightmare and author of *The Dilbert Principle*, has written, "People hate change, and with good reason. Change makes us stupider, relatively speaking. Change adds new information to the universe; information that we don't know. . . . On the other hand, change is good for the people who are causing the change. They understand the new information." Necessary change is one thing; change as a way of controlling others is something quite different.

Reorganization is a mighty weapon in the hands of an insecure manager (a surprising number of managers seem insecure). It breaks up comfortable working relationships that might, if left undisturbed, allow employees the time and peace of mind to question and evaluate the decisions of the managers. When reorganization strikes, hardly anyone is secure; personal connections are severed; paranoia increases; innovation and creativity decline. It may take months or years for the chaos to subside, perhaps just in time for the next reorganization. Frequent reorganization slows productive work and reduces its quality, both of which depend on stable, evolved relationships. Imagine, for instance, a string quartet that has played together for years, each player attuned to the slightest nuances in tempo and dynamics of the others; then a new concert manager requires that the members of the quartet switch instruments at unpredictable intervals or change places with other musicians waiting offstage.

An unusually severe bout of reorganization hit my university a few years ago. A university committee empowered by the office of the provost and led by a distinguished professor in the social sciences terrorized the five campuses in New Brunswick for months, gathering "information" and threatening major reorganization. Eventually a report was delivered to the provost. We waited anxiously, not knowing what it said, devising strategies of response to the scenarios we imagined. It was like living under the wavering ax of a drunken executioner. Then the blow fell. Without warning, the provost's office itself was abolished by the president. The reorganizers were reorganized. Naturally, the committee's report vanished without a trace, and all of our strategies were totally inadequate to meet the new disruptions. A few weeks later, I called one of the administrators who had survived in the wreckage of the former provost's office. "I'm keeping a low profile," he confided quietly, "so that I don't get my head chopped off. Those of us left here move cautiously because we don't want to slip in the pools of blood on the floor."

In government, as well as business and academia, reorganization creates permanent, institutionalized instability. In addition to the usual internal political reorganizations associated with the nearly continuous manipulations of budgets and power structures, there are external reorganizing forces, including movements for term limits, constitutional changes, and privatization. Although some of these changes can be beneficial, the sum total is not, and there is never enough time between reorganizations to see what has worked and what hasn't.

The rapidity of the electronic transactions that have replaced older ways of doing business is another force that opposes evolution. Social evolution is slow—not as slow as biological evolution, but a lot slower than the frenetic, sound-bitten, microsecond-conscious blur that we exist in now. It takes a long time to get to know people, to work with them effectively and smoothly—a long time to learn what they can and can't do, will or won't do, what turns them on and what turns them off, when they can be trusted and when they can't be trusted, when to smile and when to frown. There is no time for all that now. There is no time for social evolution, which remains stubbornly resistant to any efforts to make it move faster.

For example, ordinary letters are slower than faxes. There are times, especially in the case of routine communications that face a deadline, when a fax seems the only way to get the job done. (Of course, before faxes some deadlines were more forgiving; for those that were not, we had to learn to allow enough time to meet them. Technology has a way of creating its own self-justifying urgency.) But the speed of a fax has serious drawbacks; when I send a fax, I need to ask myself whether I can afford to work so quickly. Having the time for reflection is not a luxury and is not dispensable. A decision that seemed unambiguous when it was made may appear quite different after a night's sleep. In such cases, a rapid exchange of faxes will help to promote bad choices and the social disruption that follows.

Just as technology-fostered speed (including the speed of introduction of new technologies) disrupts social evolution—the development of stable institutions and relationships—so does technology-fostered isolation. I have already discussed the isolating effect of e-mail and the Internet; the impact of this isolation on the fabric of institutions and communities can be devastating. It is hard to imagine how the process of social evolution can work when connections among people are limited to the narrowest of information channels. Businesses may be affected more than communities, as the workplace is fractured into a collection of individually staffed home workstations and far-flung employees communicating with each other by e-mail, cell phones, and occasional video conferences. While this kind of atomized and shifting arrangement is most likely to be found in the service sector of business, manufacturing is also prone to other kinds of technology-related upheavals. Factories are now far more portable or disposable than they once were. Increasingly, when their plant moves to a different state or country, workers have to choose between losing their jobs and abandoning their homes, friends, bowling clubs, music groups, favorite hiking trails, trustworthy auto mechanics, neighborhood shops, and children's schools to follow their workplace to its new location.

Social evolution, like its biological counterpart, is a process of great depth and subtlety. Its best end products—lasting institutions, durable friendships, stable communities, accumulated wisdom, and gentle and productive cooperation—are among the highest achievements of our human existence. They help us live in a mutually beneficial way with the other products of biological evolution, which we collectively call nature, and they shield us from nature's occasional violence. In similar fashion, social evolution lets us live fulfilling lives with one another and can shield us from the violence that wells up again and again in every human society.

But the Second War of Evolution is upon us. Incessant disruption, superficial contacts, and personal isolation are making social evolution impossible. We are like brightly colored bits of glass inside a giant kaleidoscope. Oh, look at the beautiful pattern we make. Twist, it changes. Oh, look at this one. It's not so nice. Twist, it changes again. Twist. Twist. Don't look anymore, just hang on.

Better yet, let's get out of the kaleidoscope if we still can. Find others and slowly make our own patterns with them. Evolve. Teach. Persist. Live.

WRITING

The semester is over, my grade rosters have been handed to the registrar, and I am home, writing sporadically and glancing through a book by P. G. Wodehouse, whose incomparable prose, as elegant as it is funny, is always very soothing.

The thirty students in Conservation Ecology were very good this year. Most of them were informed, outdoors-loving, committed, hardworking people who cared for each other and seemed to enjoy the class. With a few exceptions, however, they can't write English, and I am still recovering from the effects of reading the term papers upon which their grades were largely based.

For twenty-five years, I have been assigning a major term paper in lieu ("in loo," as one student wrote) of an exam. I used to get eighteen- to forty-page papers, acceptably or even nicely written; now the students struggle to reach fourteen pages with the help of triple spacing, margins you could drive a bus along, and type sizes usually reserved for the visually disabled. And the writing!

The first mistake the students are making is to use the "spell-checkers" of their computer software as a substitute for proofreading. The results are papers in which all the words are spelled correctly, even if they are not the right words. Reading these spell-checked papers can be like trying to translate from Spanish to English based on the assumption that words that are spelled the same in the two languages have the same meaning. (This can lead to some confusion if, for example, the Spanish words are *sin, cabal,* or *saber.*) In one of the papers, I had to erase a long, marginal comment I had written when I realized that *illicit* was meant to be *elicit.* In another, I had trouble with the philosophical implications of the word *modal,* used throughout the text, until I turned it into *model,* which made more sense.

Then there were words such as *begum* ("a Muslim lady of high rank," my dictionary told me), which didn't seem to fit easily into the context of a paper on the genetics of coyotes, and which didn't have an obvious substitute. When I came across words like this, I just left them: it was easier to imagine Muslim ladies of high rank dealing with coyote genetics than it was to find the intended word beginning with "b" or, alternatively, ending with "m."

Complicating the difficulties brought on by the spell-checkers is a new system of placing commas, a system whose secret logic has not been divulged to me. It may be a flaw in my training, but I find it difficult and terribly time-consuming to read a, paper in, which, the commas have, been placed in a seemingly, random, manner. If a student then further complicates the writing by the sentence structure and scrambling, and not only is it inverted and, because, of all of the, parenthetical phrases, and especially using, nouns as adjectives like desert fish habitat loss (DFHL) with so many acronyms, maybe, five or, ten, to learn and forget for each, paper and with an organization based on the idea that if you, repeat, a sentence enough with a few small changes, each, time then you have organized your thoughts so you see the affect I mean effect it has on me and why after reading the papers I wanted to kick the dog except because of his teeth which are very, large.   The paragraphing:
too.

When I finished grading the last paper, the dog was asleep on my foot. Getting up quietly and carefully, I went outside to plant potatoes in the back garden with Sam. I made the furrow; Sam placed the potatoes at nine-inch intervals and covered them up with soil. His brother Jon was disdainfully shooting baskets in front of the house, not interested in what we were doing, but Sam's adolescence has taken a different turn; it doesn't seem to keep him from enjoying potato-planting with his father—yet. By the time we finished, I was calm enough to go into the house and tell my wife about the papers.

Joan listened to my lament, echoed it with some acerbic comments about the term papers she was grading for her course on global ecosystems, and said, "You realize what's happening, don't you? They're writing the way they talk. They don't know the difference between written and spoken language—that's part of the problem."

"Why don't they know the difference?"

"Because they don't get practice writing anymore. To make matters worse, they don't read either, so they don't get to see how other people write. That's why they don't know the difference between "affect" and "effect"; they sound the same when they're talking, which is all they know how to do."

"That's true," I answered. "Most of my students are seniors, and they tell me this is the first term paper they have written. They've never written anything longer than three or four pages."

The next day, I was still thinking about it. What, I wondered, are they doing in elementary or high school while they are not writing? Well, the answer to that is pretty simple; they are being tested, or, more likely, being prepared for being tested. To survive in the modern world of education, it is absolutely necessary to be able to do well on standardized tests, and no matter what the makers of these tests say, the best way to do well is to prepare early (third grade, for example) to learn the different forms of standardized questions, to memorize (but never use) standardized lists of words, and to work through many sample, standardized problems. My students may not be able to write a good term paper, but by God they can solve word analogies. Ask them: UNTRUTHFUL is to MENDACIOUSNESS as CIRCUMSPECT is to (a) FLAMBOYANT, (b) CONCISENESS, (c) CAUTION, (d) AMBIVALENCE, or (e) MASOCHISM. They'll get it right away. This analogy is based on the one cited by University of California president Richard C. Atkinson in a speech, given on February 18, 2001, recommending that his university drop the SAT I test from its admissions requirements.

How, I mused, have the educational testers, thousands of educational psychologists, statisticians, mathematicians, computer scientists, school administrators, and countless other bureaucrats managed to pull off this massive swindle while arousing only a feeble outcry of protest? How have these "professionals" convinced us that the tests are necessary and that they don't themselves interfere with and misdirect the process of education? Part of the answer must be our infatuation with anything that smacks of science and statistics. But there is more to it; the testing industry has done a wonderful job of hiding the assumptions that underlie its tests, some of the most dubious assumptions one can find in a democratic society.

The first assumption of standardized testing is that for any question there is nearly always a clear and evident right answer amidst a sea of wrong answers. Failing this, there is at least a right answer that is demonstrably stronger and better than other right answers.

The second assumption is that the meanings of words are fixed and straightforward. These meanings are not shaded or altered by context, nor should they be used differently by different people.

The third assumption is that complex ideas and subtle nuances of meaning can be reduced to brief questions with short answers that can be graded by a computer. Even written essays on these tests can be evaluated objectively and satisfactorily by poorly paid graders whose work output is itself being timed by the management of the testing company.

The fourth assumption, one of the most dangerous, is that a test score can reveal the testee's present worth, promise as a future student, and his or her level of success in life itself, years later, after formal education is done.

There is a final assumption, the master assumption behind all the others. It is that written language can be abstracted from daily life without doing it any damage. What I mean is that the testers, professionals to the core, have helped to take written language away from my students, who are mere amateurs; they have separated it from life and have made it a disembodied part of what we mistakenly call the educational process. Considering the testing assumptions that the students have grown up with—that there are always evident right answers, that context doesn't matter, that complexity reduces to simplicity, that objective tests describe the present and predict the future—no wonder they have trouble writing (and thinking) about the ecological problems of the real world.

Writing gains strength from frequent practice, but only when inspired by the events of daily living. For instance, in a short story, I came across a description of someone as "weedy" in appearance. Looking this up in my dictionary, I find that it means "tall and thin; lanky and weak-looking; of poor physique." But as a gardener who knows weeds from experience, I can give the definition added meaning. True, weeds are often tall, thin, and unattractive; they are also fast-growing, persistent, and, for all of their poor physique, surprisingly tough. The people who first used "weedy" as an adjective must have known these things—they were putting part of their lives into the word, thus making their language a bridge between nature and mind.

The writing of most of my students' term papers shows that the bridge is out; the reality of their language is a static and contrived virtual reality, consisting of disembodied images without true life, without sustaining passion. In no way does it reflect the world of their active, outdoor being. Even the punctuation and paragraphing are dead and disconnected. A comma should be nothing more than a place to take a quick breath without interrupting a thought; a dash is a deeper breath; a paragraph is a pause that allows you to look around at the landscape and see where you have been and where you are going. When writing works properly, its very words and structure should relate us to our world—assuming that we have a real world to relate to.

If written language is no longer an organic outgrowth of people's lives and environments how can it be used to communicate the urgent problems of those lives and environments and how can it help in the search for practical and lasting solutions? It won't do to say that we will simply become a postliterate, oral/electronic culture where communication is merely rapid speech made visible (for example, e-mail); this kind of language is not serving us any better than our written words. It is as disconnected from nature and the sources of human permanence as is writing.

But it does not do to dwell on these things for too long at a stretch. So if you will excuse me, I want to go outside and see whether the chicken wire fencing is keeping the rabbit away from my beets and radicchio.

THREE

DEADLY ECONOMICS

The way . . . to deliver ourselves from our own destructiveness, is to quit using our technological capability as the reference point and standard of our economic life. We will instead have to measure our economy by the health of the ecosystems and human communities where we do our work.

Wendell Berry, *Life Is A Miracle*

Money is a thing which can be owed in principle to anybody; it is the same whoever you owe it to. . . . Not many of our duties are really of this impersonal kind; the attempt to commute other sorts of duty into money is a notorious form of evasion.

Mary Midgley, *Evolution as a Religion*

AFFLUENCE AND AUSTERITY

A funny thing happened to me on the road to affluence. I have talked about it with others often enough to discover that it is a common happening and an important one. But before you read any further, you should know that what you are about to encounter is presented as a middle-class problem—nobody dies in an unheated garret, nobody is thrown out on the street with their meager belongings, nobody is in prison as a birthright, nobody is caught in a drug runner's crossfire. For all its bourgeois setting, however, my story is about the aspirations of our whole society; directly or indirectly it affects us all.

What happened is that as I got more affluent, I didn't feel any richer. Affluence, according to my Random House dictionary, is an "abundance of money, property, and other material goods." Setting aside quibbles about what constitutes abundance, there is no doubt that I now have more money, property, and other material goods than I did, say, twenty-five years ago—especially if books count as material goods. So why don't I feel wealthy? I can best explain by starting at what was for me the beginning.

I grew up in a nice house in one of the better residential neighborhoods of Passaic, New Jersey, a city composed mostly of immigrants and their children and grandchildren. The town of Clifton, a more prosperous community, began one block over; you could tell you were leaving Passaic because there was a sign. (It was not until I went to camp in Maine that I realized that in other parts of the country there can be open space between towns.) Our house had wood siding, two stories plus an attic and basement, three bedrooms, and a garage in back. It had cost

my parents $6,600 in the late 1930s—a good deal of money for a strug-
gling young doctor who was helping to support his bankrupted parents.

The property, narrow and moderately deep, amounted to about one-
sixth of an acre. For the first eight or ten years of my life, there was a vacant lot next door. I remembered patches of milkweed inhabited by big,
tiger-striped caterpillars that turned into monarch butterflies, even in a
jar. I also remember my feeling of loss when the milkweed was torn up
and replaced by an ugly yellow house. I could walk my affectionate but
dim-witted beagle to the park nearby where there was a pond with a wa-
terwheel at the outlet, and, once, a dozen newly hatched, brownish-
black snapping turtles scuttling across the grass to the water.

Whenever I stayed home sick from school, I got to see the daily events
of the household. Time never dragged; there was always somebody coming to the door. The side door, next to the kitchen, got the most traffic. I
made a fort out of the heavy, white-painted, hardwood kitchen chairs and
watched the arrivals from my protected vantage point. First was the milk-
man from Sisco Dairies, wanting to know how many bottles of pasteur-
ized, unhomogenized whole milk to deliver, and whether we needed any
light or heavy cream. Then the truck from Dugan Bakery pulled up: did
we need a coffee cake, hard rolls with poppy seeds, or Danish pastry? The
soda man brought cases of seltzer water, black cherry, and the fluorescent
orange-colored soda to which I was addicted. In the late morning or early
afternoon, Mr. Cohen, the fruit and vegetable peddler, showed up.

"What's good today?" my mother would ask.

"I got beautiful broccoli—and cheap," he answered.

My mother looked suspicious. "All right, let me see it." Then Mr.
Cohen would walk to his truck and bring back a head of broccoli and
maybe a cauliflower (which I hated). Haggling would begin; there were
more trips back and forth to the truck; eventually, fruit, vegetables, and
money would change hands.

In my early years, the iceman would arrive every few days, carrying
a big block of ice with his wonderful, hooked tongs. By the time I was in
third or fourth grade, a gas refrigerator replaced the icebox, and the iceman stopped coming. That a gas flame should cool things off seemed unnatural to me. It still does.

Sometimes the side doorbell would ring and a shabbily dressed man
with a stubble of beard would stand at the door.

"How do you do, ma'am," he would say to my mother, and maybe I would get a nod.

"Just a minute and I'll fix you something," my mother answered. Then she would make a hefty sandwich of whatever leftovers we had from the previous night's dinner—I remember pot roast—and would pour a glass of milk. The man sat outside on the side steps to eat, and later knocked on the door to return the plate and glass. I noticed an etiquette to these encounters: my mother never asked any questions, the men were always polite and appreciative without being demonstrative, and they expected to eat outside. Tramps were always traveling. I don't think I ever saw the same one twice.

The front door was used by the postman, who dropped the mail through a little slot with a hinged, elaborately carved iron cover. He came around ten o'clock and again, for the afternoon delivery, at three. When I was old enough to write letters, I learned that first-class mail to other parts of northern New Jersey took one or two days to arrive. Other than the postman, the front door saw a stream of visitors, aunts, uncles, and friends who came just to talk or to consult my mother about the various community service organizations that she helped to direct. Outside in the street there was still more activity. Every few weeks, on different schedules, the junk man, the rag man, and the knife-and-scissors-sharpening man would drive slowly by ringing their bells, pausing now and then so that people could have time to come out of their houses and call to them.

If my mother had an errand or appointment that took her away from home when I was sick, Josephine, who did the cleaning and helped with the cooking, could take care of me and tell stories about working in factories and about rural life in North Carolina. Josephine received a high salary for those days, and my father gave free medical care to her and her family, as well as any medicines and X-rays they needed. My mother occasionally complained to my father about the salary, but he saw a great many patients a day at $2 for an office visit, and we could afford the expense.

On some days, my father would take me with him to his office, where I would read comic books and talk with the patients. If it was a Sunday when the office was closed, I could watch my father make out the bills. On some, he would write "discount" and reduce the bill by a third or a half or more. "Why did you do that?" I asked. "Because they can't afford to pay more," my father replied.

Why tell about these things? I guess it has been long brewing, but what set it off was watching a woman make a phone call. As I was driving through New Brunswick I saw her on a street corner holding a telephone receiver, trailing a cord, to her ear. At first I thought it must be a call on a police or fire emergency phone, the kind attached to utility poles; then I saw, next to her, a small plastic wind screen with the telephone company logo on it. This little open box was passing itself off as a phone booth! What happened to the real booths where you could go inside and close the folding door? A biological analogy popped into my head: the phone booth has become vestigial, like the human appendix. Yes, I know that the new "booths" have been around for a long time, but I had adapted to them unconsciously, and only a sudden quirk of memory made me aware of the substitution. The point is that our affluent society can no longer afford to provide real outdoor phone booths, with doors, to shield callers from wind, traffic noise, and garlic-eating eavesdroppers. True, those who have cell phones can go into their cars and use them there, but not everybody has a cell phone, and not everybody has a car.

I began to think about some of the other things we can not afford. Two mail deliveries a day (soon we will be down to five or fewer per week), and ordinary first class mail that arrives the day after it is sent. Furniture made out of solid wood, not chips glued together. Useful services that people need, where they live. Affordable helpers in the home who can work with dignity and for fair salaries and benefits (although exploitation is hardly a modern invention). Medical care individually designed with each patient's total well-being in mind. A house to live in when one is starting a family and doesn't have a big income. The time for casual visits with friends. The peace of mind and security to open one's kitchen door to a hungry stranger.

Some features of contemporary life are wonderful improvements over the way things were when I was a child. There is, for example, more justice in today's America, even if there is still not enough, and there is a new understanding of inequality. Nostalgia can be a deadly trap if it is nothing more than a by-product of aging, but nostalgia is useful and necessary when it reminds us of better ways of living that are gone. I would not want us to return to the world of my childhood even if we could. I do want us to reconsider, in the full light of our experience, how we are going to define affluence in the world we will live in tomorrow.

I think about this when I am standing in a department store heaped with digital cameras, pocket television sets, electric can openers, video games, and cell phones, but with no salespeople in sight. These material goods do not make us affluent; they are a smokescreen to hide our increasing isolation and exploitation by an economy dominated by corporate forces outside our local communities.

But the smokescreen is dissolving; soon we will not be able to afford even the palliative glitter that is much of what the world's industry and technology are able to supply us. A country whose affluence is all produced by very poor people in other lands is in a perilously precarious condition. When I was a child, I remember hearing on the radio the postwar British speaking of austerity and the end of empire. They were very honest about it; even then their frankness impressed me. Now austerity is nearly upon us, too, spurred on by international trade with its massive, socially irresponsible movements of capital; by a "high" technology that gives us inventions without regard for their final cost and consequences; and by the dwindling supply of the cheap oil that supports all this. Unlike the British, we don't speak publicly about austerity. It is coming just the same.

What will Americans do when austerity exposes our latent poverty? The rage of a country packed with the adult equivalent of spoiled, frightened children is not pleasant to imagine. I expect denial, terminal indulgence, and random, often self-destructive violence. I hope also for something better. America is a tough society: innovative, resilient, and still full of humor. There are many people, old and young, who recognize true affluence and who are already spreading its seeds throughout our communities. Soon they will have their chance to demonstrate the richness that comes from a simpler, more responsible way of life.

DURABLE GOODS

I have been reading about the Lenape. It started as a kind of penance. Every day I pick up the newspaper and learn about different ethnic peoples who have been driven from their homeland by some other ethnic peoples or have been simply exterminated in place, a more lasting way of getting rid of them. One morning, while eating oatmeal and skimming over the news of the latest genocide in *The New York Times*, a thought occurred to me: Aren't I lucky that this isn't happening in New Jersey? Not that it couldn't, but at the moment it seems quite unlikely. Why not? I wondered. Is it our democratic government that protects us? Our melting pot society? Then another answer came to mind. New Jersey has had its ethnic cleansing already, a very comprehensive one—the destruction of the Lenape.

As a child I was taught in school about the Lenni Lenape Indians, who once lived in New Jersey and then didn't. The teaching never went much farther than that; it was pretty superficial compared with what my children learned about Indians (but not about the Lenni Lenape) in elementary school. So I decided to remedy the defect in my education and bought the only book about the Lenni Lenape I could find, a scholarly paperback by Herbert C. Kraft called *The Lenape: Archaeology, History, and Ethnography*. I bought it out of guilt—as I said, a penance. Not guilt for what was done to the native people of New Jersey, which wasn't my fault or even the fault of my ancestors, who arrived in America in the late nineteenth century, themselves victims of persecution. The guilt was because of a feeling that to live in the New Jersey–New York–eastern Pennsylvania region—that is, Lenapehoking—without knowing anything about its prior residents was disrespectful; it was a dishonoring, even an obliteration of their memory. Because of its penitential and scholarly nature, I expected my reading to be dull; surprisingly, the book turned out to be fascinating.

Names are important. I learned in Kraft's introduction that the name "Lenni Lenape" is as redundant as it sounds. (When I was in third grade I thought that "Lenni Lenape" was a silly name, and that prevented me from thinking of the tribe in a respectful way.) "Lenape," in the Unami dialect of the Delaware language, means something like "the real people," "the common people," or "the ordinary people." As Kraft points out, "Lenni Lenape" would mean "the common, ordinary people." "Lenape" is sufficient. I also noted that Kraft uses the term *Indian* without any apparent qualms. My few Indian acquaintances do also. One of them becomes angry when someone calls him a Native American; it is, he thinks, a euphemism favored by repressive government bureaucrats and their ilk.

But perhaps it doesn't matter what tribal name we use. After all, nobody, including the descendants of the original Indian residents, has the faintest idea what the various groups of Indians living in Lenapehoking called themselves before the arrival of the first Europeans. Even their languages have faded away; when Kraft wrote his book in 1986, there were only about ten people left who were fluent in either the Unami (southern) or Munsee (northern) languages, and all of those speakers were old. According to New Jersey historian John Quinn, William Penn, the founder of Pennsylvania, was an admirer of Munsee. In 1682, Penn described the language as "lofty, yet narrow, like the Hebrew. . . . One word serveth in the place of three, and the rest are supplied by the understanding of the hearer. I have made it my business to understand it, that I might not want an interpreter on any occasion."

What, then, remains of the original Lenape, apart from a widely dispersed and variously acculturated group of people who can claim one or more ancestors who lived in Lenapehoking before Giovanni da Verrazano first sighted it in 1524? Most of the tangible remains, especially from the earlier and longer preagricultural phase of Lenape existence, are those tools of daily life that do not rot or disintegrate after centuries or millennia of burial in moist, mostly acidic soils. The list is shockingly short: spearpoints, scrapers, hammers, axes, drills, pestles, mortar stones, atlatl weights, bola stones, netsinkers, pendants, pipes, and a few other artifacts, all made of stone. Later on, there were arrowheads, pottery, and a few kinds of implements and beads, some made of copper. That's about it.

The reason why I find this list shockingly short is because of the contrast with the durable artifacts of my own life. Not that I am excessively devoted to material objects, but almost everyone living today in an industrial nation is surrounded by the accumulated products of many centuries of human ingenuity, especially from these product-dominated times.

Without leaving the desk chair in my six- by seven-foot home office (a converted closet), I listed, before I got bored, fifty different metal, glass, and plastic items that would probably survive burial in soil for a few thousand years. I cheated a little because I counted the pre-Columbian stone axe, spearpoint, and pestle that I have placed on my desk to provide inspiration. Other items on the list include a glass paperweight, a brass doorknob, a stained-glass butterfly made by Jon, a triaxial milli-gauss meter (for measuring electric fields like the ones under power lines), some computer disks, and the fountain pen that I am using to write these words. I haven't even looked in the places where most of my durable goods are: in the kitchen, bedrooms, living room, dining room, bathrooms, basement, or—it would take a week to total its contents—the garage. Just one room of my house holds more species of long-lasting objects than does the entire, pitiful collection of spearpoints and scrapers that is the visible legacy of the Lenape nation.

What does their paucity of durable goods say about the lives led by the Lenape? Does it mean that their existence was dull, instinctual, animal-like—on a lower plane than ours? Is it possible to be fully human and intellectually alive without wheels, pulleys, or writing implements? Does the mind's potential develop in an environment with such a meager and primitive set of fabricated objects? The advocates of our own materialistic economics and technology would certainly say no.

The landscape itself is silent on these questions. The Lenape left behind no pyramids, no giant stone effigies standing among the oaks and hickories on the Kittatinny Ridge overlooking the Delaware, no remains of passive solar housing aligned to the first rays of the summer solstice sunrise across the Raritan. What they did leave behind, besides the names Kittatinny and Raritan, Kraft tells us, is the ghostly echo of a reputation, a whisper briefly noted by the self-centered Europeans who displaced and scattered them. The Lenape "enjoyed a noble status" among their neighbors—friends and enemies—before we came. The tribes around them, the Nanticoke, Conoy, Shawnee, and Mahican, "claimed a close connection" with them and called them the "grandfathers." Theirs was a social genius that resulted in achievements we moderns only experience in our utopian dreams. The Lenape had a "fiercely egalitarian" society characterized by no authoritarian leaders and a power-ful sense of community. They cherished the environment that supported them, cared for orphans, and respected the elderly. And during most of the long years when they occupied the land, they appear to have lived in peace.

Since the glaciers melted off the face of northern Lenapehoking about ten thousand years ago, sea level has been rising. We believe that many of the early coastal settlements, fields, and hunting grounds of the Lenape were covered by the waves. An imperceptibly slow process at most times, the cumulative change would have been apparent during an unusually high storm tide, as the waters rolled farther inland than they had ever done in living memory. How did these prehistoric Indians respond to such catastrophic events? What did they think, what did they say as they stood on high ground under the night sky and listened to the sounds of the sea pounding at their frail huts, sapling frames covered with chestnut bark? Did they address the Great Spirit, Kishelëmukong, "He who cre-ates us by his thoughts," in his place in the twelfth and highest heaven, with words like these?

> He who makes the Pleiades
> and Orion,
> And turns deepest darkness
> into morning,
> And darkens the day into
> night;
> Who calls for the waters of
> the sea,
> And pours them out upon
> the face of the earth . . .

The words are those of Amos of Tekoa, a herdsman and tender of fig trees, first of the Hebrew prophets, who lived in what students of the Lenape have named the Early Woodland Period. I have no trouble thinking that similar words, albeit with different constellations cited, might have been uttered by some long-forgotten Lenape living at the same time among a people not yet swept up in the tide of what we call history.

It would be a great mistake to assume that the Lenape can be encompassed and defined by the durable goods they left behind. It is the sort of mistake made by a materialistic culture, a culture preoccupied with illusions, a culture confused about what is lasting and what is fleeting, confused about what is greater and what is lesser, a culture whose future is no brighter now than that of the Lenape on that death-foreboding morning when the first Europeans stepped ashore on the beaches of Lenapehoking.

SPENDING OUR CAPITAL

The rich man's wealth is his strong city;

The ruin of the poor is their poverty.

Proverbs 10:15

A deeply felt aversion to spending accumulated capital is an ancient part of the heritage of most societies. Although my father was a doctor, not a businessman, he taught this to me. He had lived through the bankruptcy of his own parents during the Great Depression, watching as they gradually sacrificed the inventory of their store in Passaic to keep the family in food and clothing. On one side of this store, my grandfather sold records, phonographs, and sewing machines and repaired the appliances that he sold; on the other side, my grandmother, a brilliant dress designer, prepared bridal gowns for customers who came from as far away as New York City. She would dress the brides on the wedding day, too, and was celebrated for her ability to make the plainest bride look beautiful. But as the depression wore on, business fell off, the customers stopped coming from New York, and the stock of goods dwindled away. There was no choice but to close the store; my grandparents' livelihood was gone forever. Later, my father and the oldest of his four brothers made it a priority to pay their parents' creditors in full, a decision that entailed sacrifices in itself. The lesson was handed down to me: you can spend your earned income and any interest you may have received, providing you first set aside a portion to increase your savings; but *never* spend the principal, your capital, except as an act of final desperation.

To most of us, capital is associated with business, yet the habit of preserving capital and handing it on to the next generation started, I am pretty sure, not as an economic or financial practice, but as an agricultural one. In Neolithic societies, it must have begun when farming replaced hunting and gathering as the main source of food. From Anatolia to North Africa to Peru, the staple grains of wheat, rice, corn, millet, oats, barley, and rye, the legumes such as peas and beans, and other vegetables from squashes to radishes, were almost all annual crops. Every year, our ancestors needed new seeds and tubers for planting; without new seed they would have starved. So every year a portion of the harvest, including seeds from the finest and most productive plants, was saved for the next planting. This seed would have been considered inviolable; to eat it was to risk starvation and death. The idea of preserving capital, the saved-up goods and assets upon which future earnings depend, is a simple and inevitable extension of the ancient rule of saving a part of the harvest for use in planting next year. The story of Joseph in Genesis is a biblical parable that stresses this life-saving mandate to preserve capital. Foreseeing seven years of abundance to be followed by seven years of famine, Joseph receives Pharaoh's permission to require that extra portions of food produced in Egypt during the years of plenty be saved in central storehouses against the lean years to come.

Even in this profligate century, there have been those who have paid the ultimate price rather than consume the necessities of future generations. The Russian plant biologist Nikolai Vavilov, father of modern crop plant protection, organized and directed the gathering and classification of viable samples of crop seeds from all over the world in his Institute of Plant Industry in Leningrad (St. Petersburg). By 1941, as Vavilov, chief preserver of the world's agricultural diversity, was dying of malnutrition somewhere in the Siberian Gulag, a victim of Stalin's and Lysenko's hatred of evolutionary biologists, and as the German armies were approaching Leningrad, more than 187,000 varieties of wheat, rice, peas, and other crops from all over the world were represented in the institute's gene bank. Vavilov's deputies rushed to preserve the priceless collection, preparing duplicates of the most valuable samples, evacuating some to the east, and regrowing many of the older seeds on a small plot of land near the front lines. As starvation grew rampant in blockaded Leningrad, the institute's scientists did their best to protect the seeds from pillagers, rats, and exploding shells. Dimitry Ivanov, a rice specialist, died of starvation while preserving the several thousand packs of rice in his keeping. Georgy Kriyer, in charge of the medicinal plants, L. M. Rodina, a keeper of the oat collection, Alexander Stuchkin, a specialist in groundnuts, and other extraordinary men and women slowly starved to death rather than eat the agricultural capital that surrounded them. The continuing, heroic efforts of those who survived saved the gene bank for a largely oblivious posterity.

Oblivious because saving capital has gone out of style in the postwar era. We have developed a flair for squandering that would be the envy of even the most accomplished wastrel. Show us something that our children and grandchildren will need for their survival, and we will find a way to spend it or ruin it, even if the spending and ruining require much effort. Some of these efforts are remarkably ingenious; take, for example, the draping of a massive government building or an entire island in petroleum-based plastic sheeting as—I suppose some would call it—an artistic statement about light and form. Is one a grinch for asking how many barrels of our children's vanishing oil supply were used in this bizarre public display of reckless waste? Is it far-fetched to see a similar heedlessness in the political effort to lower the federal tax on what is already the most dangerously underpriced gasoline in the world? Is it unreasonable to question the decisions to sell the nation's emergency oil and helium reserves merely to hold down the prices of these limited commodities? I naively thought that the whole point of supply-and-demand capitalism was that high prices would serve to limit the consumption of scarce goods.

One of the insidious causes of our pervasive thoughtlessness is that stocks of capital are no longer as easy to assess as they were when all we needed to know was how much seed was stored in the shed or how many sewing machines and phonographs were still on the shelves. Now, the car is in the driveway in Massachusetts or Montana, but the oil wells are in Kuwait, Nigeria, Saudi Arabia, and the Gulf of Mexico. Or the commodities themselves, such as helium, are so remote from our daily experience that most of us have no reason ever to give them a thought.

Our waste of capital is also related to the problem of exponential growth. When the practice of saving capital evolved, population and consumption were stable or were growing slowly by today's standards. In the second half of the twentieth century, a rapid increase in both occurred. This has confounded our efforts to save capital in three related ways that were described in 1978 by an American physicist, Albert A. Bartlett. Bartlett applied a well-known formula concerning growth rates to the question of how rapidly resources are being used up. From this basic and simple equation, he developed principles that he thought every educated person should know and illustrated them with a series of vivid examples. First, if we assume an annual growth rate in consumption of 7 percent, the amount of a resource used during the next ten years will equal the total of all preceding consumption of that resource in history. Our conventional thinking, based on millennia of slow growth or no growth, does not prepare us for such a dramatic depletion of our resources. *Growth makes us underestimate our future needs for capital.*

Second, with exponential growth, the time at which the diminishing stocks of a resource still appear sufficient and the time at which the resource runs out occur surprisingly close together. Imagine, says Bartlett, a single bacterium put in a bottle at 11:00 P.M. The bacterium doubles every minute. By midnight, the bottle is full. When was the bottle only half full? The chilling answer: at one minute to midnight. By the time the more perceptive and activist bacteria first notice that they are running out of space and call a public meeting, it may already be too late to do anything about it. *Growth makes us exaggerate the time we have left in which to take action to save capital.*

And third, counter to all of our intuitions and to the claims of oil companies and politicians, in exponential growth the lifespan of a finite resource is only weakly related to the size of the remaining stock. In other words, the number of oil fields or copper deposits left to be discovered makes little difference in estimating the date when reserves of oil and copper will be exhausted. To return to Bartlett's bacteria, let us assume that at 11:59 P.M. an enterprising bacterium discovers three empty bottles off the north coast of Alaska. This quadruples the space resources available to the crowded population. The bacteria emit a collective sight of relief. But how long will this new space last? The calculation is very simple—only two more minutes. How much oil is in the earth? It doesn't matter. Assume, said Bartlett, that the entire volume of the earth is oil and we can extract every drop. At the then-current 7.04 percent yearly growth rate in oil consumption, an entire earth-full of oil would last just 342 years! *Growth makes a mockery of our hope that undiscovered resources of capital will save us.*

The list of dwindling capital resources is not confined to obvious commodities such as oil, coal, helium, and scarce metals; it includes items that we used to take for granted, including soil, fresh water, clean air, open space, trees, and fish in the sea. Moreover, there is another kind of capital than conventional natural resources, namely human capital, and we are wasting that, too. By human capital I mean the skills in agriculture, crafts, academic learning, mechanics, the arts, and all the other necessary human occupations. I also mean the community-fostering skills of judgment, patience, consideration, and knowledge. Human capital, like crop varieties but unlike iron ore in the earth or intact forests, does not preserve itself automatically until it is used or damaged. It demands an ongoing, active, proficient effort to prevent its inexorable decay. It can disappear in one generation—ten generations of work may not bring it back. Only a stable society can provide the continuity needed to preserve human capital.

Parents, teachers, and elders are the prime caretakers of human capital, and the low esteem that modern society accords to these civilization-preserving people helps explain why human capital is now melting away. In a world dominated by huge corporations and global trade, the need of humans and their societies for continuity has become subservient to a nonhuman economic demand for cheap production, with its ever-changing technologies and job definitions. In the present economic environment, stability is considered stagnation, and human capital is not considered at all.

In recent years, the job placement advisers have begun to say to graduating seniors, "You will each have to get used to the idea that you should not expect to have one job that lasts your lifetime—you will have to be flexible about what you do and where you live." In other words, stability is an undeserved luxury; traditions of wisdom and workmanship are not worth the effort to acquire, let alone pass on; skill is short-lived, or at best utterly generic; and saving and planning are the quaint activities of characters in history books and old novels. We have come a long way from the farmer who took comfort during the dark winter night in knowing that the seeds for spring planting were safely stored.

Today, even as "conservative values" are blasted from loudspeakers everywhere, the venerable custom of protecting our assets is rejected and vilified. Everything is topsy-turvy as self-proclaimed conservative capitalists lead the assault against the conservation of capital. Here is an evil that no easy remedies will erase. Saving capital is hard work, and our quick-profit culture imposes extra social and economic penalties on those who undertake the effort. We are not yet asked to make the extreme sacrifice of a Vavilov or an Ivanov, but pulling back from the seductive attractions of corporate culture—the supermalls, the television, the deadening conveniences, electronic and otherwise—that are eating away at our true and only wealth like a cancer, calls for a quiet heroism and a daring creativity of its own. Heroism in choosing the more difficult and sometimes more expensive path; creativity in finding new ways of saving capital in an environment that despises prudence and loathes the future. Our task has been described concisely and elegantly in Proverbs 13:22: A good man leaves an inheritance to his children's children.

SAVING BY SELLING

Since the fall of Soviet communism, capitalism—perhaps we should say big business—has been prescribed more than ever as the cure for any ill, including the decline of nature. A large part of the natural world has been damaged or destroyed by unregulated commerce. Now, various groups of conservationists are trying to save some of the most spectacular remnants of nature, both species and ecosystems—with more commerce. The idea of turning the tables and using the methods of exploiters to prevent more serious exploitation is an interesting one, but the risks are high and not everyone in the business of saving by selling seems to have given much thought to them.

I first came across commercial conservation in one of its earliest and most dubious forms, the farming of sea turtles, a phoenixlike enterprise that, despite its chronic inability to turn a profit, always springs up renewed from the ashes of its last bankruptcy. Because of its biological and economic complexity, the farming of sea turtles nicely illustrates many of the problems inherent in the commercialization of consevation and is worth examining in some detail.

There are seven species of sea turtles in the world's oceans, ranging in size from the approximately 100-pound Atlantic and Pacific ridleys and the hawksbills, to the huge leatherbacks, which can weigh as much as 1200 pounds and possibly more. All except the loggerhead nest primarily on tropical or subtropical beaches, although some venture far into cold waters in between nestings. The eggs of every species are prized as food and in Latin America are considered an aphrodisiac. The leather is used for expensive shoes; the hawksbill provides shell for jewelry; and the hawksbill, loggerhead, and especially the green are taken for their meat, cartilage (for soup) and oil (for cosmetics). Indeed, the green turtle has been described as the world's most valuable reptile. Not surprisingly, nearly all sea turtle populations have been seriously depleted, and the great majority of the nesting populations that existed 150 years ago are now extinct.

The theory of sea turtle farming goes something like this: green turtles (and other sea turtles) are in trouble because nesting females and their eggs have been taken from the nesting beaches in huge numbers, the turtles are caught in the waters adjacent to the nesting beaches and at their distant feeding grounds, and in recent decades they have been drowned by the thousands in the trawls of shrimping vessels. If unchecked, this depletion will lead to total extinction of these slow-growing animals. How much better, the argument runs, to remove an expendable percentage of eggs from the nesting beaches, hatch them, raise the hatchlings to market size, and sell them. This will displace wild-caught animals from the market because of the easy, guaranteed availability of the farm-raised turtles and the higher, more uniform quality of the domestic product. As a bonus, we can set aside a few breeders to supply eggs in the future, thus creating a closed system with no net drain on wild populations. It sounds good, especially to those who know nothing about sea turtles and to those who have a vested interest in the operation.

From the beginning, sea turtle farming was strongly opposed by Archie Carr, the preeminent turtle biologist of the twentieth century, and by a sizable group of his students and colleagues, including me. The reasons we gave for this opposition, and which we still give thirty years later, are rooted in a mix of biology and economics. Here are a few of the more important ones.

The first problem is the unavoidable expense of coping with the biology of a nondomesticated species, a frequent worry for conservation entrepreneurs. Sea turtles mature slowly. Some green turtle populations may take nearly a half-century to reach sexual maturity in the wild; they are not exactly broiler chickens. To speed this up, captive turtles must be fed expensive food enriched with animal protein, something akin to trout chow. Feeding efficiency is low because the pellets dissolve or are washed away before the turtles can eat them.

Holding tanks are expensive to build, and sea water is expensive to circulate and filter. Fencing off bays or inlets for impoundments instead of using tanks doesn't work very well; the fences get holes in them or are destroyed by storms, or the impoundments become heavily polluted. In both tanks and impoundments, unnaturally crowded turtles get sick from a variety of bacterial, fungal, parasitic, and viral pathogens. At the National Marine Fisheries Service's Galveston, Texas, facility for "head-starting" Atlantic (Kemp's) ridley turtles, twenty-seven major categories of disease were noted among hatchlings. At the Cayman Turtle Farm on Grand Cayman Island, West Indies, flagship of the (originally) commercial operations, one outbreak of chlamydiosis is reported to have caused a die-off that cost the company hundreds of thousands of dollars. Other turtle-rearing facilities around the world have mortalities that typically range between 50 and 75 percent, and sometimes higher.

There are problems concerning reproduction, the primary measure of the success of any conservation plan. Keeping a breeding stock of sea turtles for a large operation is economically (and biologically) impractical. I once calculated that it would take 293 turtles to supply just 50,000 eggs per year, even assuming that all biological obstacles to producing fertile eggs in captivity are overcome — an unwarranted assumption. The space required for these very large breeders and the cost of feeding and maintaining them make this notion of breeding stock a manifest absurdity. Only the Cayman Turtle Farm has made a serious effort to supply its eggs from its own captive-reared breeders; the hatching rates from eggs laid by these breeders, averaged over a twenty-year period, was just 6 percent, compared with 70 to 90 percent at well-managed hatcheries at natural nesting beaches. Obviously something, whether diet, accelerated growth rate, or captivity itself, is terribly wrong.

So turtle farms are really turtle ranches, according to the accepted definition of the Convention on International Trade in Endangered Species, CITES for short. They are not true farms, because they are not self-maintaining—they can't produce all the eggs they need. To survive, green turtle ranches must take eggs from the dwindling wild populations every year. The eggs they take are referred to as "doomed" eggs, ones that would not hatch if left where they were laid on the beach—a conveniently slippery and subjective notion. As an added sop to conservationists, turtle ranchers have released, with much fanfare, some ranch-raised yearling turtles; a small number of yearlings is supposedly an adequate replacement for thousands of eggs and hatchlings in the wild, most of which would not survive to maturity. Because of the complex life cycle and migrations of sea turtles, however, there is no evidence that the majority of released yearlings ever find and join their wild relatives, or that they produce viable offspring. After more than thirty years of releases of tens of thousands of head-started turtles of at least four species in a variety of programs, from 1996 to 1999, only nine head-started individuals have been documented to nest in the wild.

The third problem is one of market stimulation, a problem that is common to most kinds of profit-making conservation. To sell their products and keep up the necessary cash flow, turtle ranches have to create a stable demand for turtle meat, soup, oil, and leather, and for dried, varnished yearlings or hatchlings preserved in lucite blocks or any other turtle products people will buy. Because of the high costs of ranching sea turtles, all these products must be priced as luxury items. At the same time, the high prices and especially the developed, stable markets make it extremely attractive to poach wild turtles and find ways of slipping them into commerce. Poaching is always cheaper than ranching; therefore, there may not be any way to protect an endangered species once the world market is stirred up, even in the name of conservation. The power of global demand erodes all safeguards. The problem is one of scale; the world market is just too big for most natural systems. The commercial ranching of green turtles inevitably brings us around again on the downward spiral, a little closer to the extinction of the remaining populations. By no stretch of the imagination is this conservation. Nor is it good business. After a series of owners and no history of profits, I have heard that the Cayman Turtle Farm has been taken over by the British government, I assume because of its value as an important tourist attraction and source of jobs in the Cayman Islands, not because of its sales of turtle products.

All of the problems in the ranching of sea turtles are generated by an interaction between biology and economics. Few other commercial species that can be raised in captivity have the intractable biology of sea turtles. Do similar difficulties exist in other examples of species conservation as a business venture, with or without the introduction of global marketing? Some scientists say yes. Valerius Geist, an eminent Canadian wildlife biologist, has repeatedly warned against the dangers of selling wildlife or wildlife products. Among the many hazards he mentions are genetic pollution of wild stocks and the spread of diseases such as tuberculosis and spongiform encephalopathy (mad cow disease, known as Creutzfeldt-Jakob disease in humans) to wild populations of buffalo, elk, and deer by animals that escape from game ranches. Although it would be difficult to prove, the recent dramatic increase in spongiform encephalopathy among game animals in some western states may indicate that Geist was right.

But responsible proponents of conservation marketing also exist. Gordon Grigg, an Australian zoologist and conservation biologist, advocates the regulated market hunting of wild kangaroos as a profit-making incentive to sheep ranchers to hunt kangaroos and raise fewer sheep. Sheep, Grigg points out, compete with the native kangaroos for food and damage the sensitive vegetation of the Australian outback with their alien hard hooves and destructive grazing habits. Crocodiles, iguanas, and other wild species have been ranched, with varied results. Because the species, environments, and human interactions are always different, each case has to be examined on its own merits.

The commercial exploitation of entire ecosystems, as opposed to single species, can again be very problematic as a conservation technique, for many of the same reasons. For years, the preservation of the New Jersey Pine Barrens has been justified by some conservationists as a way of saving the enormous aquifer of pure, fresh water that lies beneath. But if that water were ever to be exploited on a large scale, the water table would fall, and even a small drop would cause extensive changes in the ecosystem, including major loss of wetlands.

"Ecotourism" is also being widely advocated by conservationists, with incomplete knowledge of its effects in most cases. There is no question, however, that watching whales from boats and lions and antelopes from Land Rovers, and bringing large numbers of visitors to rain forests and deserts can have adverse effects on the species and habitats being visited. Similarly, the campaign to sell nuts and fruits from tropical rain forests has preceded any serious attempt to find out whether the exploited forests can sustain these worldwide sales, however well intentioned. Botanists Rodolfo Vasquez and Alwyn Gentry have reported that the collection of forest-harvested fruits by natives for sale in regional markets, a flourishing trade, may not be sustainable in the highly diverse Amazonian forests around Iquitos, Peru, where there are hundreds of kinds of trees per acre. There are simply not enough seeds left in the forest after fruit collection to replenish the populations of fruit trees. On the other hand, the botanist Charles Peters and three of his colleagues who studied Brazilian and Peruvian Amazonian forests of low tree diversity, primarily swamp and flood-plain forests with only one or two dominant tree species, came to a different conclusion: these kinds of forests may be able to sustain "market-oriented extraction of fruits" indefinitely.

Increasingly, there are highly publicized efforts to investigate and develop the commercial value of traditional medicines extracted from tropical rain forests. In one study, in Belize, botanists Michael Balick and Robert Mendelsohn found that the strictly local use of products gathered from small harvested clearings, with long intervals between cuttings, does not seem to damage the forest or the harvested species. This is encouraging, yet it would be wrong to extrapolate from this finding and assume that no harm would come from commercial extraction of medicines on a grander scale. As in the case of sea turtle ranching, local forests may not be able to survive the demands of the world market.

Specific biological knowledge is needed to determine the feasibility of each venture in commercial conservation. On this biological basis, some projects, such as sea turtle ranching, ought to be ruled out from the start. Others, such as the use of forests to provide traditional medicines, may prove feasible. Nevertheless, even in these latter cases the economic scale of the operation is likely to be limiting; there will be few schemes of commercial conservation that supply a globalized market without destroying the species and ecosystems they purport to save.

Finally, even if we do find an occasional example of saving by selling that is not demonstrably destructive, we should ask ourselves whether it is worth the moral cost of the effort. Can a lasting human relationship with the environment be based on the premises that conservation must always pay for itself, that there is nothing of lasting, intrinsic value left in this commercial world, and that the world can be preserved without rich people having to make sacrifices?

HOT SPOTS AND THE
GLOBALIZATION OF CONSERVATION

Eight friends were seated around the dining room table, two bulging photo albums at the ready, about to start a round of that curious adult version of Show and Tell known as Our Summer Vacation.

Donna and Stanley had begun their album with pictures of Stanley delivering a lecture to clinical psychologists at the Catholic University of Campinas, in the state of São Paulo, Brazil. This was followed by page after page of lush, colorful South American scenery: snapshots of the mighty Iguazú Falls taken from the river below; towering walls of green vegetation along the Rio Paraná; a thick growth of native trees overtaking the exotic plantings left behind by the celebrated botanist Moises Bertoni at the remains of his agricultural experiment station in the Paraguayan province of Alto Paraná. Then there were pictures of Stanley and Donna with their Brazilian hosts, flanked by stately palms; pictures at their friends' beautiful beach house in Ubatuba; pictures of costumed samba dancers; and more pictures of scarlet and blue macaws, gaudy toucans, purple bougainvilleas, brilliant butterflies, and the rich magnificence of one of the last remaining patches of Brazil's Atlantic coastal forest.

The album provided a breathtaking display of natural and human exuberance painted in a profusion of vivid reds, blues, yellows, and, above all, greens. I could almost smell the myriad fragrances borne on the tropical breeze and hear one of my favorite sounds, the chattering of parrots far above in the treetops.

As the pages turned, the phrase "hot spots" came into my mind, that graphic term for those patches of exceptionally high biodiversity—many of them in the tropics—that receive the lion's share of attention from conservation biologists and others dedicated to the preservation of endangered species. I thought of the conservation argument that says if we want to preserve life, look first at the places where, acre for acre, the most kinds of life exist—where an acre can yield a hundred species of trees instead of a handful, and more species of insects than most of us can imagine. This understandable preoccupation with hot spots is a kind of biological version of Sutton's Law, named after the accomplished bank robber, Willie Sutton. When Sutton was once asked why he robbed banks, he gave the now-classic reply, "Because that's where the money is." The biological "money" is in the hot spots, it seems.

My thoughts drifted back to an article published in *Science* in January 1997 by Princeton University's Andy Dobson and three co-authors, in which they analyzed the geographic distribution of endangered species in the United States. Because their analysis of the data indicated that many endangered species of both plants and animals are found localized in a few hot spots of diversity—Hawaii, southern California and Arizona, southern Appalachia, and parts of the southeastern coastal region—they claimed that our endangered species can be protected most efficiently by spending conservation efforts and funds in these few key areas. I had been bothered by this article when it first appeared, and was still bothered by it as I looked at Stanley's and Donna's pictures. But then it was Joan's and my turn to show vacation photos.

Our album was open to a picture of a small, twin-engined Otter airplane disembarking thirteen warmly dressed people and waterproof bags, canvas Duluth packs, and partly disassembled canoes onto a flat expanse of gravel. Everything in the foreground lies in the gloomy shadow of early twilight, but a low, treeless hill in the background, yellow-brown in color, is brilliantly illuminated by the summer sun. Nothing in the picture is green. Not a single plant is in evidence. "It's very bleak," Judy remarked. "It looks like a desert."

It had felt bleak, I remembered, especially when the little plane, its tundra tires churning the lichen-covered stones, took off, climbed steeply out of the narrow Soper River Valley and disappeared over the rounded hills of the Meta Incognita Plateau headed back to Iqaluit, once known as Frobisher Bay. We had been left at the upper reaches of the only canoeable river on Canada's vast Baffin Island, north of Hudson Strait, in the eastern part of the newly designated Inuit territory of Nunavut. At eight in the evening, the Arctic summer sun was still fairly high in the sky but hidden behind a hill. A chill wind was blowing. Several members of our group glimpsed a wolf on the other side of the river. We found a partially sheltered spot off the gravel bar and pitched tents in the stunted vegetation while our Maine guides, Randy and Dwayne, cooked dinner on two Coleman stoves. That night we were awakened briefly by the sun shining on our tent at 3:00 A.M.

The next page of the album revives my spirits; it shows us two-thirds of the way up Mount Joy, with the Soper River winding through the valley below. "Everything is so brown," somebody comments. "Wait a bit," I respond, knowing what's coming in a few pages. Meanwhile, here is a shot of Mark showing off a hefty arctic char, kin to trout and salmon and the one species of fish in the freshwater part of the Soper or any other river on Baffin Island—a landmass that stretches farther than the distance from New York to Chicago. Char are carnivorous; the young eat the tiny shrimp and mosquito larvae that are among the few invertebrates in the river, while the older char eat the younger ones as well. Not exactly an ichthyological hot spot, I muse as the page turns.

Next, Joan and I are shown running the first big set of rapids, which Randy has dubbed "Gatorade." What the Canadian geographer J. Dewey Soper called it we have no way of knowing. He doesn't give it a name in the first published description of the river named after him, which appeared in 1933. The page turns again, and suddenly everyone crowds around to get a better view. There are the thundering, 95-foot falls of the Cascade River, a tributary of the Soper, with the giant block of ice miraculously clinging to the adjacent cliff. In the foreground is a patch of "mountain cranberry" dotted with last summer's tasty red berries. The only kinds of trees that grow on Baffin, shrubby willows and birches, most less than ten to fifteen inches high, cover the lower slopes of the valley of the Cascade. "How beautiful," Donna says, and I nod; when we were there I had told Joan that this was the most beautiful place I had ever seen. Soper, the first outsider to set foot in this valley with its tumultuous river and falls, called it "a fascinating commingling of savagery and softness." Another photo shows a dry upper hillside above the valley. Close-up photos reveal the mosses growing between the stones, with every rock sporting a multicolored array of encrusting lichens.

The page turns and there we are hiking back from the falls to our third campsite. We see the tundra ponds surrounded by low, green vegetation, looking, if you squint a bit, like an aerial photo of a New Jersey salt marsh in June. I thought I saw tiny fish in one of those ponds, but when I brought Dwayne, a fisheries biologist, to look, he pronounced them shrimp. Then I got down close to them and noticed that they were swimming on their backs, just like the fairy shrimp that my teacher, Archie Carr, once took me to see in a temporary pond in north Florida.

Farther down the Soper River, at the fourth campsite, Joan and I stalked a caribou. This is easier than it sounds, in spite of the absence of cover; perhaps they don't see well, or perhaps they are unafraid and curious. The picture shows the caribou, looking too small for its ridiculously large antlers, browsing on plants in a gully about thirty feet from us. We could hear the crackling noises it made as it chewed. When we got back to camp, Randy, who tracks and studies bears for the state of Maine, pointed out a sandy patch where the ever-present caribou tracks were overlaid with the surprisingly big prints of a solitary wolf. The photos of the tracks came out well, but I was sorry that I hadn't listened to Joan and put my Swiss Army knife down next to them to give an indication of their size.

A few miles away, Randy directed us inland to the only two stands of real trees, willows, on Baffin Island. They grow in two adjacent patches of several acres each. Some of the willows are as tall as twelve feet. Dewey Soper found and photographed a similar patch, now dead, not far away in the valley. In this part of the Canadian Arctic, the treeline is four hundred miles to the south, in northern Quebec; Soper was very excited to come across actual trees that stood up from the ground rather than forming low, spreading mats. Joan thought that the location of the patches in hollows next to sheltering hillsides allows deep snow to accumulate, protecting the willows in winter from the freezing and drying Arctic winds. A photo shows Joan in the midst of the amazing forest. Only a few of the skinny trees are taller than her head. Another picture shows me beside one of two six-foot willows that Joan and I found next to a small stream in a steep-sided little valley several miles away. It is possible that no one else has ever seen these two trees. You can see that I am as proud as if I had discovered a giant sequoia anchored in the permafrost. (This is not as far-fetched as it seems. My friend John Tedrow, an expert on Arctic soils, has seen big logs, perhaps dawn redwood or larch, embedded in the Beaufort gravels of Prince Patrick Island's polar desert, far to the north and west of Baffin, where they probably grew several million years ago when the climate was warm.)

More exclamations from the people around the table. We have come to the pictures of arctic flowers blooming in what Soper described as "unrivalled loveliness . . . in the briefest, though the fairest summer season in the world." Joan has marked each plant in the album with its common and scientific name. Here is a lowland wet spot covered with eight-inch stalks of the cottongrass sedge, *Eriophorum angustifolium*, its cottony tufts of plumed seeds waving in the breeze. Not in the photo are the billions of seeds drifting by in the sky above the ridge tops, making the light refract and sparkle as if the air were filled with tiny diamonds. On the next page are the ubiquitous showy purple flowers of arctic willow herb, *Chamerion (Epilobium) latifolium*, and underneath on the page, a rounded cluster of yellow mountain saxifrage, *Saxifraga aizoides*, looking like an arrangement in a florist's shop. Then comes a picture of mountain avens, *Dryas integrifolia*, with Joan's Swiss Army knife looming over it; and here is a clump of white heather, *Cassiope tetragona*, that has nearly finished flowering in late July—a few white blossoms can still be seen among the seed capsules.

The major study of Baffin vegetation identifies only 189 species of higher plants on the entire island, an area larger than the British Isles. In our seven days on the river, Joan found and keyed out 54 species, a significant part of the flora. She usually spent an hour or so before dinner, looking through her hand lens at each of the day's plant specimens spread out in her lap, while others in the group explored, rested, or gathered around Tom while he sang and played the guitar.

The album ends with pictures of some of the beautiful and vivacious Inuit children who accompanied us everywhere on the last day of our trip while we were in Kimmirut, formerly Lake Harbour, on the southern coast. There is also a picture of three skulls of polar bears—killed in self-defense—resting on top of a shed; and a final view of the Meta Incognita Plateau, brown, lifeless-looking, and laced with meltwater streams and narrow patches of snow, as we flew over it on our way back to Iqaluit.

Joan closed the book. I could see that the people around the table were moved by what they had seen, but there was no way that Joan and I could convey the depth of our feelings about this place, at once so vast and unforgiving, yet so intimate and welcoming, so poor in species diversity, yet so vibrant with life. What we encountered in Baffin Island, and what made it magical, was that increasingly rare quality of nature not yet vanquished, an amazing and nearly pristine ecosystem still in place, still working. It is not a hot spot, it is an evolutionary pinnacle of a different sort, nature fully equal to the terrible rigors of a harsh climate.

Why, I wondered, has species diversity become a god to conservationists? The wealth of species that adorns the tropics is one of the greatest glories of this planet; it must be protected. But wealth of species is only one way of reading nature; there are many others. A preoccupation with hot spots can absorb too much of our energy and concern, leaving little for the rest of the natural world. Unspoiled places like the valley of the Soper River, poor in species as they are, have their own kind of grandeur and glory, a living presence that surely makes an equally powerful claim on our care and affection.

Nevertheless, the idea of hot spots persists and has now captured the public's attention. The British conservationist Norman Myers, who first used the term back in the 1980s, published a paper about it, with four coauthors, in the February 24, 2000 issue of *Nature*. The twenty-five principal hot spots in the world, mostly tropical, contain "as many as 44% of all species of vascular plants and 35% of all species in four vertebrate groups" in an area that amounts to "only 1.4% of the land surface of the Earth," Myers claimed. The word "only" is an important one; it leads to the key idea in the paper, which in turn is probably the reason why *Nature* gave such prominence to this article about hot spots.

The idea is simple and is based on a fact that is hard to impeach: there are limited funds available for conservation in the world. Given this fact, say Myers and his colleagues, "How can we protect the most species per dollar invested?" The answer is simple, we need to implement "a 'silver bullet' strategy [a specific remedy precisely targeted to a particular problem] on the part of conservation planners, focusing on these hotspots in proportion to their share of the world's species at risk." Stuart Pimm, an ecologist at Columbia University who coauthored *Nature's* commentary on the Myers paper, was also quoted in a March 14, 2000, *New York Times* article on hot spots written by William K. Stevens. "They're probably going to capture pretty much 30 to 50 percent of everything," said Pimm, referring to a silver bullet campaign to save hot spots. The Myers study was done, he said, with "extraordinary care and numeric detail"; the paper is "one of the most remarkably optimistic statements about biodiversity that we have ever seen." Russell Mittermeier, the president of Conservation International, an expert on primate conservation and second author of the Myers paper, was also mentioned in the *Times* article. Mittermeier referred to their statement in the paper that $500 million spent annually would "go far towards safeguarding the hotspots." This is an amount of money that is only twice the cost of a single Pathfinder mission to Mars, and vastly less than the $300 billion annual cost of protecting worldwide biodiversity that had been previously estimated. Five hundred million dollars, Mittermeier observed to the *Times*, is an achievable goal.

So for $500 million per year, not even the cost of one Stealth bomber, we can begin to protect as much as one-third to two-fifths of the earth's biodiversity. There are individuals who could part with $500 million without any appreciable change in lifestyle; there are multinational corporations for which $500 million is less than their advertising budget; there are governments for which $500 million is an insignificant fraction of their annual expenditures.

It is no wonder that for Myers, Mittermeier, and Pimm this news is cause for great optimism, and their hopeful message has been conveyed by the newspapers and public radio to many hundreds of thousands of people. Theirs is a feel-good message because it means that countless rare species might be saved. Not only will these complex ecosystems continue to function, but somewhere inside them may well be living the species of plant, microorganism, or even animal that will end the scourge of cancer, provide the next generation of antibiotics, give us new miracle pesticides, inspire more inventions such as Velcro, and maybe—if we're lucky—cure baldness.

But there is another, more important, reason why the message of hot spots makes us feel good; it reassures us that we, in the richest and most powerful of countries, don't have to make any fundamental changes in the way we live our daily lives, the way we behave toward poorer nations, and the way we treat the other 98.6 percent of the land (not to mention the two-thirds of the planet that is ocean) in order to save biodiversity. With a trivial expenditure to safeguard the hot spots, an expenditure that is much less than 1 percent of the annual gross global product, we can keep on with business as usual.

Toward the end of the Myers article, the authors discuss conservation funding. They claim that during the ten years between 1989 and 1999, nongovernmental organizations such as the MacArthur Foundation, Conservation International, and the World Wildlife Fund have spent an average of $40 million per year on hot spots conservation. This, they assert, "is only a tiny fraction of the amount spent per year on biodiversity conservation by governments and international agencies, these funds being assigned mainly to across-the-board activities rather than the concentrated efforts advocated here." Although they describe their "silver bullet" strategy of concentrating on hot spots as complementary to other conservation activities, their description of these other efforts as "scattergun . . . seeking to be many things to many threatened species," implies that the hot spots strategy is a better alternative, not a complement. We must set conservation "priorities," they claim. "How can we support the most species at the least cost?" they ask.

I take it as established fact that there is now a mass extinction taking place as a result of human activities; that the bulk of the species extinctions are taking place in the tropics, because that is where most of the species are; that there are limited funds available for international conservation programs; that to the extent that these programs work, a conservation dollar spent in a hot spot is likely to save more species than one spent elsewhere; and that if we don't save biodiversity now, it will not regenerate during the future course of human history. Nevertheless, I am unhappy with the hot spots concept—both with the way it was presented by Myers and his colleagues, and with its subsequent popular development.

There is an obvious problem with hot spot–style conservation, in addition to others that are more subtle. A look at the map of hot spots presented by Myers reveals a striking feature. Most of the terrestrial world is not hot spots; there are many large countries, such as Argentina, Mongolia, Canada, Zaire, and Poland, that have no hot spots in them. If the Myers team was simply pointing out places that are especially rich in biodiversity and are deserving candidates for conservation, perhaps we could ignore this distribution, but it wrote of limited conservation resources and the need for setting priorities. This clearly implies that we have to choose between the Philippines and Canada, and the Philippines wins. Or Brazil's Atlantic forest versus Argentina, with Brazil winning. The tropics versus the temperate and boreal regions, with the tropics winning. Don't we have enough problems without making conservation a nationalistic battleground? Even within countries, this is a divisive strategy. The California floristic province is a wonderful place, but I don't think it is a good idea to divert conservation resources from the New York–New Jersey Highlands or the Maine woods to this western hot spot.

A less obvious but equally important flaw in the hot spots approach is that it treats the value of nature as a numbers game. A density of 5,000 species per square mile is better than 2,500; 10,000 is better than 5,000. I suppose that this comes from our number-obsessed, science and technology-obsessed culture, in which anything that can't be quantified is suspect. But there are other ways of looking at the things that matter, as the trip to Baffin demonstrated. There's more to nature than the number of species in it. Moreover, although the tree diversity in the tropical hot spots is much higher than it is in my native New Jersey, for example, few of those tropical trees would survive a New Jersey winter. Wonderful as the hot spots are, the rest of the world is not going to be biotically maintained by their conservation.

Before we devise global conservation solutions—if that is what we are trying to do—we ought to step back and ask what is causing the biotic impoverishment that these solutions are attempting to address. I don't mean to imply that there is only one cause; there are many, and they are intertwined. But there is one that stands out for me, globalization, which is the divorcing of business from any association with place or community. Globalization as a cause of the problem is getting obscured by globalization as a purported cure, a situation that must be avoided at all costs.

To explain what I mean, I want to look at some of the implications of the hot spots strategy. The approach I am going to take was first suggested to me by one of my graduate students, Kristi MacDonald. I was discussing the Myers paper with her, and she remarked that hot spots are the conservation equivalent of globalization. I had thought a good deal about globalization, although never in that context, and I immediately got a sense of what she meant. Then, by chance, a few days later, I came across a featured column in *The New York Times* that made it perfectly clear. It was written by Thomas Friedman, the *Times'* first apostle of globalization, on March 31, 2000. Friedman, a staunch defender of the I.M.F. and the World Bank, writing from the Venezuelan rain forest, criticized the then-forthcoming protests against those organizations in Washington. "The truth is the I.M.F. and World Bank . . . are bit players, of declining importance, in a much larger drama. . . . The real drama is on the ground in the developing world . . . [and] the real players are a diverse collection of companies, indigenous peoples, endangered species and local governments" who, he reports, are working together to save the Guyana Shield, the world's largest unbroken expanse of tropical rain forest. Friedman was there with a team from Conservation International, which is brokering a deal to put together a 13 million-acre protected area in the Venezuelan part of the Shield. The deal involves debt-for-nature swaps and the participation of Venezuela's national oil company, Petróleos de Venezuela, which owns Citgo in the United States. Why is Petróleos participating? Because, according to Friedman, they want to enhance their global environmental image. Friedman quotes Glenn Prickett, senior vice president of Conservation International, as saying, "Thanks to globalization and the Internet, power is now much more diffused, global companies are now much more exposed, and organizations like ours much better positioned to offer solutions." So don't just protest, Friedman counseled the Washington protestors, go out and do something positive to help globalization save the environment.

I fervently hope that the 13 million-acre protected area and home of the Ye'kwana people comes into being. However I am not very sanguine about the chances that real, durable conservation is going to happen there, regardless of what is put on paper, and I am even less happy with the projection of this deal as a model of effective, globalized conservation. First, debt-for-nature swaps have not been overwhelmingly successful; they depend on too many uncontrollable factors, including the value of the national currency, which has to remain low without becoming absolutely worthless for a long period of time if the swap is to succeed. Political factors concerning land ownership also complicate and can damage debt-for-nature swaps, as Conservation International has discovered elsewhere in South America. I also wonder whether Petróleos, which has enlisted the World Trade Organization to undermine U.S. environmental laws, will care so much about its environmental image if its technical advisers should discover oil, diamonds, gold, or uranium in the reserve; or what will happen when a new, environmentally unfriendly government comes into power. But these are not my major concerns.

The connection of hot spots with globalization, as Friedman must realize, is that the *hot spots are almost always somewhere else.* That doesn't seem to bother him, but it bothers me. When I was in Baffin Island, I understood the tremendous importance of conserving its magnificent ecosystems, but I never thought that they should be saved instead of the ecosystems of my home, New Jersey. After his brief, manicured visit to the Guyana Shield, Friedman came back to New York, whose endangered species and threatened habitats are not displayed on his monitor. That's the nice thing about hot spots for Friedman and many other supporters of globalization: they don't have to live in them day in and day out. Globalization, however it is rationalized by the likes of Friedman, is not the free and equal trade of goods among prosperous nations everywhere, nor will it be. It is the use by multinational corporations of cheap, semi-slave labor in poor, mostly tropical countries to grow cash crops and make goods that are sold in the rich, temperate zone countries—at least for the few more years that those countries remain rich. The economist Herman Daly and the theologian John Cobb have written in their book, *For the Common Good,* "Free traders, having freed themselves from the restraints of community at the national level and having moved into the cosmopolitan world, *which is not a community,* have effectively freed themselves of all community obligations."

The consequences of this uncoupling have been terrible, and responsible people feel guilt over the global social and environmental activities of Shell, Occidental Petroleum, W. R. Grace, and hundreds of other giant corporations, guilt enhanced by our reckless enjoyment of our transient wealth. Enter the hot spots. We can now save nature while driving our SUVs.

But it is time to admit that conservation is not just a matter of throwing dollars and experts at the primarily poor, primarily tropical places that have the most species. This is an imperialist idea; it reeks of condescension and neatly transfers responsibility from the people who have caused most of the problem—us—to those who are suffering the consequences of our acts. "We give you back a small, small part of the money we took from you—now use it to protect those pieces of your land that we designate, while we and our associates in your country continue to despoil the rest."

The alliance between multinational corporations and conservationists to protect hot spots and similar places is an unholy one. Although some temporary good may come of it, we cannot afford to let it make us more accepting of globalization, more contemptuous of the local cultures that preserved the hot spots until the onset of globalization. Here is an extract from Barbara Kingsolver's extraordinary novel, *The Poisonwood Bible*; it makes the point very well.

[The first Portugese explorers] reported that the Africans lived like kings, even wearing the fabrics of royalty: velvet, damask, and brocade. Their report was only off by a hair; the Kongo people made remarkable textiles by beating the fibrous bark of certain trees, or weaving thread from the raffia palm. From mahogany and ebony they made sculpture and furnished their homes. They smelted and forged iron ore into weapons, plowshares, flutes, and delicate jewelry. The Portugese marvelled at how efficiently the Kingdom of Kongo collected taxes and assembled its court and ministries. There was no written language, but an oral tradition so ardent that when the Catholic fathers fixed letters to the words of Kikongo, its poetry and stories poured into print with the force of a flood. . . . The Europeans were dismayed to find no commodity agriculture here. All food was consumed very near to where it was grown. And so no cities, no giant plantations, and no roads necessary for transporting produce from the one to the other. The Kingdom was held together by thousands of miles of footpaths crossing the forest, with suspension bridges of woven vines swinging gently over the rivers.

Marvelous systems like this, with the natural world and the human community completely intermingled, have an enormous, evolved complexity. They are particular, unique, and fragile to outside disturbance. Globalization is inimical to these qualities; bringing it into the conservation process is like recruiting the executioner to become a nursemaid. Not all native peoples have lived in harmony with their environment, as the ruined soils of what was ancient Mesopotamia still show, but never has the destructive side of human nature been given such scope and power as it has with globalization.

Maybe asking which areas of the world should receive the highest priority for preservation is not a good question. A better one is, "How can we live in a way that reduces the burden on species and ecosystems elsewhere?" Another good question is, "How can we live so that we protect the species and ecosystems in our own community?" Perhaps these are the same question; certainly they have the same answer.

The plant geneticist and ecologist Wes Jackson, co-founder of the Land Institute in Salina, Kansas, has increasingly turned his attention to the question of how to live successfully in a community. In his book, *Becoming Native to This Place*, he writes, "An extractive economic system to a large degree is a derivative of our perceptions and values. But it also controls our behavior. We have to loosen its hard grip on us, finger by finger. I am hopeful that a new economic system can emerge from the homecomer's effort—as a derivative of right livelihood rather than of purposeful design." More than two decades ago, Wendell Berry had the same message in *The Unsettling of America*: "The possibility of the world's health will have to be defined in the characters of persons as clearly and as urgently as the possibility of personal 'success' is now so defined. Organizations may promote this sort of forbearance and care, but they cannot provide it."

How do we make personal forbearance and care a force in the world? Wes Jackson says that one of the most effective ways for this to come about "would be for our universities to assume the awesome responsibility to both validate and educate those who want to be homecomers— not necessarily to go home but to go somewhere and dig in and begin the long search and experiment to become native." An admirable hope, but I am sure that Wes knows that the modern university is the last place where this is going to happen. All research colleges and universities that I have seen are willing captives of globalization, always chasing money, power, and "world-class" scholarship in some other part of the world, always contemptuous of things that have the taint of the merely local and merely useful about them. The distinguished historian Page Smith, founding provost of the University of California at Santa Cruz, wrote in his book *Killing the Spirit,* "In traditional societies, institutions may have a comparatively long and useful life before they become impediments to the ends they were established to achieve. This process seems to be far more rapid in 'new,' 'open,' 'democratic' societies and is certainly observable in the American university. . . . When the organizational structure of an institution of higher education is indistinguishable from that of a major corporation, the spirit dies."

Barring revolutionary upheavals, universities, except for scattered individuals in them, are not going to help. Berry has more practical suggestions, published in the *Amicus Journal* in 1992. A few of his recommendations are to buy near at hand, rather than at a distance; buy from a local store rather than a chain store; do not buy anything you do not need; do as much as you can for yourself; if you can't do it for yourself, see if a neighbor can do it for you; buy food that is locally grown; grow food for yourself; take your vacations as close to home as possible.

The relationship between conservation and home has been explicitly and eloquently spelled out by the ecologist Chris Uhl, writing in the December 1998 issue of *Conservation Biology*:

> The ultimate challenge for conservation biology isn't to create more parks, manage more forests, or shepherd through more environmental legislation. Rather, the challenge is to change the way we as a society perceive the natural world. Einstein was right: "We shall require a substantially new manner of thinking if mankind is to survive." The parks and enlightened legislation will come rushing in once we get our thinking right. . . . As mundane as it may sound, for many of us the land at our doorstep provides the starting point for developing an affection for the earth, which is a necessary foundation for living respectfully within the confines of our

planet. The healing of the earth will spread outward from the land around our doorsteps.

There is something very special about local conservation, something that operates at the most profound level of human existence, something that is lacking when we think about global conservation, important as it is. Whenever I drive along the Millstone River, here in central New Jersey, I am reminded of a particular night in January 1976. Earlier that day, my nine-months-pregnant wife, Joan, who hates sitting around, came home early from work and shoveled out most of our 100-foot driveway after a snowstorm; now, at 3 A.M., we were on our way to the Princeton Medical Center for her to give birth to our second child, Jane. The snow had stopped and the stars were out. Our road ran for miles along the Millstone River, peacefully flowing under the ice past snowy fields and woods whose pin oaks, river birches, and box elders stood out stark and beautiful against the white background. The memory of the river and trees on that winter night is an indelible part of me. Whenever I see them again I feel a deep sense of gratitude that they are still there, thanks to the efforts of my state's many conservationists. Any threat to them would be a threat to my own person.

The force of such memories and associations cannot be exaggerated: the vacant weedy lot next door where at the age of five you first saw a monarch caterpillar on a milkweed leaf, the mating calls of peepers and spadefoot toads coming through your open bedroom window on a spring evening, the smell of the pine woods where you sometimes see a richly colored corn snake sunning itself on the sand road while you walk your dog. Most of us have memories like these, even if we don't know it. Local conservation has continuity; it is passed on from parent to child, friend to friend, and it has a strength that has to be seen to be believed.

If you listen to the corporate globalizers who dominate the mass media, you will get the impression that local conservation is a hopeless struggle against overwhelming forces that you had better join rather than confront. Nothing is farther from the truth. It is the globalizers who should worry—they still have a lot of damage to do, but their days are numbered, as the unmanageable techno-economic behemoth they have created collapses of its own weight.

There are valid applications of the hot spots concept. One that was flourishing thousands of years before the days of globalization was described to me by my friend Frank Lake, an American Indian and an experienced field biologist from northern California. Frank has told me that in the Indian country that he knows best, the sites that the native peoples held most sacred, and which they managed and protected with the greatest care, appear to be the places that had the highest number of species and the highest natural productivity. In Frank's words:

> My father and I would visit the high country in the summer months between the time when I was eight and thirteen years old. It was at this time I learned the importance of being a person of place and what it meant to have healthy and productive watersheds. Our visit was almost always for religious purposes. I was taught that everything in this area was sacred and that it must be respected. Respected even if I did not understand it, and that in time, if I observed what was going on around me, and I was patient, I would eventually learn to understand and value the power of place.

This use of the idea of hot spots is good biology, good conservation, and good community service. These sacred hot spots were identified and conserved by *local* people to meet *local* physical and spiritual needs, the two of which were inseparable. One cannot ask for better conservation or a more durable economics than that.

THE GINGKO AND THE STUMP

I had a call from Bill Stevens at the science desk of the *New York Times*. "I'm doing a story about an article that is about to appear in *Nature*," he said, "and I'd like to fax you a copy for your comments. The article is by Robert Costanza and twelve other economists, geographers, and ecologists. It's called 'The Value of the World's Ecosystem Services and Natural Capital.'" As soon as I heard the title, I knew that I was not likely to see eye to eye with the authors. With some reluctance, partly because Costanza is a committed and accomplished environmentalist whom I didn't like to criticize, I agreed to read and comment on the article. Soon I had it in hand.

I started out to read the article thoroughly, word by word, examining the methods, the assumptions, and the numbers in the tables with care. But it was hard to do; I kept wanting to skim. Not that it was badly written; it was clear and lucid. Not that it was aggressive or arrogant in tone; indeed it was mildly apologetic. Not that it was ill-intentioned; its goal was to demonstrate that nature in the form of intact ecosystems is worth far more to us than we currently imagine. Nor were the estimates exaggerated; as the authors claimed, they were obviously understated, giving a minimal figure of $33 trillion for the annual value of nature. This is approximately twice the value of the global gross national product, they pointed out. In other words, nature is very valuable to us, even more valuable than the goods and services that people produce.

One problem was immediately obvious. The authors did not seem to realize the dangers of what they were doing. Never mind the larger dangers to our essential humanity, but even the practical, economic dangers. Suppose, for example, we are able to increase greatly our global gross national product, perhaps to $50 trillion or $75 trillion? Why worry about nature then? Or, what if a group of antienvironmental economists calculates the global *costs* of nature's harmful activities: the floods, the droughts, the damaging winds, the insect pests, the weeds, the infectious diseases and their vectors? What if these calculated costs amount to $34 trillion?

But something else was wrong, more fundamentally wrong. I looked again at the text. Nature was reduced in this analysis to seventeen "ecosystem services and functions," including "climate regulation," "water supply," "soil formation," waste treatment," "pollination," "food production," "recreation," and "cultural." Adding the estimated values of these seventeen items gave Costanza and his coauthors their figure of $33 trillion dollars, enabling them to compare the worth of nature's services with the global gross national product. The reductionist absurdity of this Swiftian exercise first irritated, then amused me. What really got to me, however, was the feeling that the effort to assign economic values to particular functions of nature was somehow terribly misguided in a different way, even if I couldn't put my finger right on the problem.

I faxed my comments back to Bill Stevens, and a few of them were duly printed in his story. "Common sense and what little we have left of the wisdom of our ancestors tells us that if we ruin the earth we will suffer grievously," he quoted me as saying, followed by the statement, "I am afraid that I don't see much hope for a civilization so stupid that it demands a quantitative estimate of the value of its own umbilical cord." Stevens slightly softened the bluntness of my remarks by pointing out that I accepted Costanza's numbers as conservative estimates. And that, I thought, was that. But it wasn't, really — this kind of argument stays with you, stewing quietly as you go about other business. The weekend came.

On Sunday, Joan, Sam, and I bought a gingko tree. There was a large gap, right in front of our house, in the line of trees growing along the street. The city had offered to plant a tree in the empty space, but I didn't want any of their white ashes or Iranian elms shading out the incredibly productive sour cherry in the lawn next to my house, so I said no. I wanted a gingko, with its beautiful leaves, its wonderful history as a living Mesozoic fossil rediscovered in a Chinese temple garden, its high tolerance of street pollution, and its growth form compatible with the neighboring cherry tree. After trying four different nurseries, we found one with a perfect gingko: about five feet tall, so that Joan and I could move it ourselves, and with a natural gingko shape, unlike the skinny, columnar varieties that are commonly available. A grafted male, it would not be dropping any vomit-smelling fruits on the sidewalk in the fall. The nurseryman said that it was a variety known as October Gold. It cost $89.95.

On the drive home, I thought about the value of the piece of nature that we had purchased. What was it worth to us now? Clearly $89.95, because that is what we had just paid. Would we have spent $179.90? No, that would have been too expensive. Maybe, I thought, Costanza is right. We can assign values to nature; we do it all the time. But people have grown accustomed to taking nature's services for granted (like the services of housewives). Why not wake them up by showing how much these services are really worth? All right, I answered myself, show them what nature is worth, but why do it in dollars? That's the enemy's turf. There the internal argument rested, undecided.

In twenty minutes we arrived home and carefully wrestled the gingko, with its heavy, burlap-wrapped root ball, out of the station wagon to the spot where we intended to plant it. The sun illuminated its little May leaves and reflected off the plastic, $89.95 price tag. I wondered whether I should have sent those comments to Bill Stevens.

The time was 11:15. I had to drive Sam to his soccer game at 2:30. No problem. I had more than three hours to dig a hole about twice the size of the root ball, put in the gingko, and fill the spaces with dirt from the hole mixed with a couple of bags of high-quality topsoil. Three-quarters of an hour would be ample.

The spot I had chosen was a slightly depressed, circular area about two feet wide where the grass had never grown properly. On the first swing of the pick, I hit something solid that gave off a chunky sound. A few more strokes of the pick and some shoveling uncovered the problem. It was a large buried stump, probably from a silver maple, which had been there for at least the fifteen years we had owned the house. "Move the hole a few feet away," Joan suggested, but that sound advice convinced me that no other spot was suitable. Besides, I reasoned, if the stump has been in the ground so long, it was probably rotten and would be easy to remove. "Have it your own way," said Joan, and she went off to plant some flowers along the front walkway.

I started in with the pick and a solid, straight-bladed, Irish garden spade, using both to dig around the stump and to break off pieces of it. The trench around the stump deepened to only three or four inches before I hit roots and other obstructions. The pieces of stump that I was able to break loose would have embarrassed an ailing chipmunk. Time passed and I labored on, getting first hot, then exhausted. Sam and Joan took pity on me and joined in. Sam was delighted to get a chance to use the pick.

More time passed; we all grew tired. We could see four or five inches of the stump in some places, but the bottom of it remained well hidden deep in the earth. Joan went back to her flowers, and Sam and I continued, taking turns with the pick. I was now angry at the stump and at myself; most of my pick strokes were missing the mark. Then Sam said, "Dad! You're not doing anything. Why don't we try to split the stump?" Grateful for the idea, I leaned on the spade while he went off to get a sledgehammer and wedge. We worked at this for a little while and made a little more progress. Occasionally a two-inch piece of stump would break off, but the wedge got stuck frequently and had to be loosened with tiring, sideways blows of the sledge along with strenuous pick work. My watch said 1:00. At the rate we were going, it would take another four hours or more. The soccer game was still at 2:30, Sam reminded me.

Then my neighbor, David Freedman, came by. He is a practical man who was raised on an anarchist agricultural commune and has owned and run factories for the majority of his more than seventy years. "Burn it," he advised, handing me a bottle of waste gas-oil mixture and some matches. Soon the stump was blazing merrily, while I looked around nervously for the police. Every now and then I would squirt some charcoal lighter on it, and I even added some charcoal to increase the heat. By 1:45, the stump seemed only lightly singed, Sam was openly skeptical, and I remembered that the gas line from the street to our house ran only ten feet away. Lugging over the garden hose, I put the fire out. Not quite a quarter-inch of the wretched stump had been consumed. The remainder seemed fire-hardened. Wearily, I picked up the sledge while Sam positioned the wedge.

A shadow fell across the stump. I looked up. It was my new neighbor from across the street. I had forgotten his name. He was holding a black iron bar, five feet tall, with a blunt, inch-and-a-quarter-wide chisel blade at the bottom. The bar tapered to a sawed-off circular striking surface about five-eighths of an inch wide at the top. "Try this," he said.

"What is it?" I asked.

"I think they call it a shale bar. I found it in the garage when we moved in." It was very heavy, about twenty pounds. I came down hard with it on to the stump and drove it further in with a few blows of the sledge. The stump split, and the length and weight of the bar let me easily pry the two halves apart. By 2:00, the stump was lying on the sidewalk in four massive pieces. Twenty minutes later, the gingko was planted and Sam and I were off to the soccer game.

The next day, I went to our local masonry and hardware dealer to buy a shale bar of my own. There were a half-dozen of them standing in a corner along with a different type of bar that was a little taller and thinner and had a curiously expanded, pancake-like top. "I would like one of these shale bars," I said casually to the man behind the register.

"That's a pinch bar you've got there. The shale bars are the other ones."

"What's the difference?"

"The pinch bar is heavier. Believe it or not, that makes it easier to use. Our customers mostly break up concrete with them. The shale bars are for going through shale—like if you were excavating for a swimming pool. And the pinch bars are five dollars more. That'll be $34.99, plus tax."

At home, I looked at the pinch bar more closely. Sighting along the length of it, I noticed visible irregularities that made it look handmade. The label said "Rat Tail Crow Bar"—an interesting zoological hybrid—"Cotter & Co., Chicago, IL, Made in USA." On a whim, I called Cotter & Co.

"We're a distributor, not a manufacturer," said the second person I spoke with. "That bar came from the Mann Edge Tool Company, in Lewistown, Pennsylvania." I thanked her and called Mann Edge.

The man I talked with in Lewistown was able to tell me a bit more. "Those are high-carbon steel bars," he noted, "made in a rolling press. But we don't make them. We get them from the J. & H. Company, in Columbiana, Ohio. You should call them and talk with John Kephart."

When I finally got through to John Kephart, I learned that he was the "J" of J. & H., the "H" being his wife, Helen. "We're one of the few plants left in America that make those bars," he told me. "We've survived because this is a small, tightly run family operation with a low overhead and no retail sales expenses. Our furnaces are gas-fired and we recycle our solutions, so we don't have a pollution problem. Your bar was made from 1060 high-carbon steel in an open die, using a piece of equipment known as a forging roll. Our presses were mostly made in the teens on into the 1940s. We have to be able to cast parts and fix them ourselves."

As I had guessed, forging with an open die involves much hand craftsmanship and operator skill. The pinch bar starts out square in cross-section near the bottom and gradually becomes round farther up. That's hard to achieve. It takes about two years to learn how to make a bar.

"We're proud of their quality," Kephart said. "Also, we refuse to temper the ends of them, as some companies do. That would make the bar harder, but might introduce microscopic heat cracks, causing the bar to break under stress. So our bar will only last you ninety-five years, five years less than a tempered bar."

I learned from Mr. Kephart that pinch bars are mostly used for positioning railroad ties and moving machinery. From his words, it was clear that their use, like their manufacture, is an art. I enjoyed talking with him—here was an expert, I thought, whose knowledge was useful and trustworthy.

After the call, I stepped outside to admire the gingko tree and removed its $89.95 price tag. Then I went over to survey the pieces of the stump piled by the driveway. What was the value of the stump, I suddenly wondered? Not a great deal as firewood. Was there anything else? Well, it had given Sam and me a chance to work hard together—and it had made me listen to his advice. The physical exercise was surely beneficial. I had strengthened my ties to two excellent neighbors while involving them in a task that improved the neighborhood—restoring a tree to a place where one had been before. I had gained understanding of a tool and a craft that belonged to America's history, and, I hope, to its future. Yes, the stump was worth a great deal more than the purchase price of the gingko, but the actual amount would be impossible to calculate. Even nature that has been fought, dismembered, and discarded has its value. So much for attaching a dollar price to nature, I thought, as I carried my new pinch bar to its place in the garage.

THE DEATH PENALTY

The example of the American States shows that private corporations—whose rule of action is the interest of the association, not the conscience of the individual—though composed of ultra-democratic elements, may become the most dangerous enemies to rational liberty, to the moral interests of the commonwealth, to the purity of legislation and of judicial action, and to the sacredness of private rights.

George P. Marsh,
Man and Nature; or Physical Geography as Modified by Human Action

On May 22, 2000, Randolph Walker, a seventy-one-year-old career actor, was struck and killed by a double-decker tour bus while crossing a street in midtown Manhattan near the low-rent apartment complex where he lived. This sad accident—an all-too-common occurrence in crowded cities—nevertheless galvanized the New York community for reasons that should concern us all.

Mr. Walker was on his way home after an audition for a bit part in a movie called *Dummy*. He had been hopeful about getting the part; he was always hopeful. This is probably why he had been able to eke out a living as an actor for nearly twenty-five years after leaving a job in banking. He pieced together bit parts in plays, work in dinner theaters, and "under five" roles in soap operas, in which an actor is hired for one day and has less than five lines to speak. In his obituary in *The New York Times* (June 1, 2000), reporter Robin Pogrebin quotes a friend, Todd Heughens, as saying, "'There are so many actors who don't work at all, and here was a guy who supported himself as an actor. . . . So I think he took great pride in that.'" Yet he was described by his fellow actors as "noncompetitive in the extreme. . . . He told his friends about roles he was seeking so they could try to get them also," and "he was happy for them even when they won parts he had wanted."

Mr. Walker, who was six feet tall and slender and had a resonant voice, was generally cast in small character parts, such as butlers or undertakers. He loved acting and was an inspiration and mentor to younger actors, according to his agent, Michael Hartig, who was also quoted in the obituary: "'When he worked on shows, they would gravitate toward him,' Mr. Hartig said. . . . 'He always had time to talk to them, to share his knowledge and his expertise.'" He was a quiet, unassuming professional who lived for his work, did not resent his lack of fame, and left behind nothing but good reports and sadness when he died.

Ironically, it was the manner of his death that secured Mr. Walker a firestorm of public sympathy and a major obituary in the *Times*. The driver of the bus that struck him was unlicensed to drive it, and Mr. Walker had the right of way. Later, an investigation by city and state agencies of the bus company, New York Apple Tours, found a shocking history of hundreds of violations of city, state, and federal laws and regulations, and it set off a wave of outrage against the company.

Nine days after Mr. Walker was killed, Governor George Pataki ordered the State Department of Motor Vehicles to suspend the vehicle registrations of all 63 of New York Apple's buses. "We can't tolerate an operation that ignores the regulations and flouts the law," the governor said. "They should be, in my view, put out of business." Mayor Rudolph Giuliani agreed. But on June 14, a state Supreme Court judge, Phyllis Gangel-Jacob, reversed the governor's order on the grounds that it would irreparably harm Apple Tours—and the company was back in business. The city and state appealed the decision, and on December 14 a five-judge panel of the Appellate Division of the State Supreme Court reversed Judge Gangel-Jacob, temporarily banning Apple's buses from the streets again. Apple Tours appealed this judgment to the state's highest court, the Court of Appeals, and in January, 2001 that court dismissed the suit, allowing the suspension to stand. On February 8, Apple Tours surrendered its operating license and, according to its lawyer, went out of business. By that time, it had been fined more than $8 million and was selling off its assets, including about a dozen double-decker buses purchased by a division of Coach USA.

Despite these events, State Senator Thomas K. Duane, who worked with Manhattan community groups to try to shut down the company, was unwilling to declare victory. "They always have another angle," he told Sherri Day, a reporter for *The New York Times*, who published her story on February 26. "You can't trust what they say, and you can't trust what they do. We still want to have a permanent revocation of their license, because no matter what they say, we want to make sure that they'll never be able to do business in the City of New York again." Presumably, Senator Duane was gratified when the New York State Department of Transportation permanently revoked the bus company's charter in April. As reported by Ms. Day in the *Times* on April 9, a spokesman for the state agency said, "If the principals wanted to reestablish Apple Tours or another company, they would come under very close scrutiny."

A Republican governor and a Republican mayor helping to put a corporation to death for its crimes! Don't they know that for more than a century, corporations—at least the very big ones—have been immortal, shielded by our system from threat of termination? Their names may change, their officers come and go, they may merge or split, reorganize, or go bankrupt, but the corporation lives on, its "rights" intact, regardless of what it does, forever.

It wasn't always this way. Until the last third of the nineteenth century, corporate charters could be and were revoked, for cause, by the states that granted them. Even as late as 1890, Justice Finch wrote for a unanimous N.Y. Court of Appeals in the case of the North River Sugar Refining Company, "The judgment sought against the defendant is one of corporate death. . . . The life of a corporation is, indeed, less than that of the humblest citizen." Where corporate transgression "threatens the welfare of the people," Judge Finch continued, "they may summon the offender to answer for the abuse of its franchise or the violation of its corporate duty." The judgment of the court was to order the immediate dissolution of the company.

The U.S. Constitution doesn't mention corporations. The idea of corporate immortality in U.S. law dates back to 1819, with the decision of U.S. Chief Justice Marshall in the case of Trustees of Dartmouth College vs. Woodward. Marshall spoke of "a perpetual succession of individuals . . . capable of acting . . . like one immortal being." But as the chief justice also said in that decision, a corporation is "a mere creature of law, it possesses only those properties which the charter of its creation confers upon it." Presumably, if a corporation violates its charter, the law can unmake it. Its immortality is conditional.

All that has changed. For the largest corporations, charters have become little more than legal formalities, and corporations have rights and powers in court greater than those of any individual citizen. People all over the United States have learned this to their sorrow, for example while fighting in court to stop the encroachments of cellular towers and supermalls on unspoiled landscapes in their communities. As George Marsh wrote more than a century ago, "More than one American State is literally governed by unprincipled corporations, which not only defy the legislative power, but have, too often, corrupted even the administration of justice. The tremendous power of these associations is . . . [due partly] to an old legal superstition—fostered by the decision of the Supreme Court . . . in the famous Dartmouth College case—in regard to the sacredness of corporate prerogatives. There is no good reason why private rights derived from God and the very constitution of society should be less respected than privileges granted by legislatures."

With the emergence of the World Trade Organization (WTO), the little chance that courts and legislatures will control the largest corporations has all but disappeared. A three-member panel of the WTO, meeting in secret, has invalidated a section of the U.S. Clean Air Act on behalf of the Venezuelan oil industry. The WTO has also ruled against the part of the Endangered Species Act that protects sea turtles from needless death in the nets of foreign shrimpers. And it has sanctioned the European Union for its ban on the hormone-laced beef produced by U.S. agribusiness. Thanks in part to the unrelievedly pro-multinational rulings of the WTO, it is our laws and regulations that are being terminated, not corporations.

Did the governor realize what he might be starting when he sought the death of New York Apple Tours? Most viable political candidates, Republican and Democrat alike, currently support the death penalty for people. Is it possible that the leaders of the major parties can be persuaded that we need the death penalty for corporations? After all, there are corporations which, on the strength of the evidence, are more deserving of lethal injection than the residents of Texas's Death Row.

But the death penalty should never be imposed lightly, even for a corporate entity whose life is worth "less than that of the humblest citizen." Innocent people's jobs are at stake; creditors may lose significant amounts of money when the corporation is shut down. Also, the act of closing certain kinds of manufacturing plants, mines, and waste depots can generate new risks to the environment and public health, creating the need for years, maybe centuries, of expensive monitoring and maintenance of the abandoned facilities. Nevertheless, some corporations have histories so depraved and products so deadly that the public welfare would seem to demand the death penalty.

In a country ruled by laws, matters of life and death are resolved in the courts, after legislatures have defined the crimes and penalties. What would a trial of a corporation accused of capital crimes be like? Once the trial has started, certain critical questions will inevitably be raised. Did the highest executives of the company know that their products were damaging people and the environment, and did they conceal that knowledge? The tobacco industry—which has given us a preview of such trials—has been shown to have knowingly and covertly laced its cigarettes with dangerous, addictive chemicals. Similarly, in August 2000, the president of Mitsubishi Motors acknowledged that his company had been concealing customer complaints about faulty fuel tanks, clutches, crankshafts, and brakes since 1977. The tobacco industry's actions have resulted in thousands of deaths, but in the case of Mitsubishi there is no evidence to date that the defects have caused any serious or fatal accidents.

Meanwhile, it was revealed in August and September 2000, that defective Bridgestone-Firestone tires on Ford Explorers *have* caused many deaths—knowledge that both Bridgestone and Ford seem to have covered up, or, at best, ignored. In a trial, would this fatal outcome make a difference? Equally important, would prosecutors be allowed to admit evidence concerning the environmental damage caused by the Explorer's heavy gas consumption, excess exhaust emissions, and off-road capabilities? Corporate violence against people and violence against nature often go together.

A question bound to be raised at trial is whether there is a long history of ruthless abuse of corporate power. Some of the largest chemical companies have been devastating environments and killing people in and out of their factories for many decades. Senior executives of the giant German chemical and pharmaceutical combine, I. G. Farben, were convicted at the Nuremberg Trials in 1948 for complicity in such corporate crimes as "plunder and spoliation" and "slavery and mass murder." The sentences meted out for these crimes were short, and two of the executives who served time in prison later went on to become CEOs of I. G.-successor companies. These successor corporations, among the most powerful in the world, are still flourishing and still causing harm to people and environments. For example, the Bayer Corporation, as late as 2001, refused to stop selling the antibiotics known as fluoroquinolones for use in chickens and turkeys, a lucrative market. This was despite the strong protests of the U.S. Centers for Disease Control and the Food and Drug Administration's Center for Veterinary Medicine, because of mounting evidence that feeding fluoroquinolones to poultry is causing widespread bacterial antibiotic resistance that can make these Cipro-type drugs useless for treating dangerous infections of humans in hospitals.

What about crimes so enormous that they cannot be conclusively proven until after civilization and the biosphere have suffered irreparable damage, and perhaps not even then? For example, multinationals, especially some of the interlinked oil, chemical, and automobile corporations, are vigorously fighting every meaningful effort to slow global warming. Meanwhile, the ice in Antarctica appears to be melting, and oceans and corporate profits are on the rise. Many of these most serious corporate "crimes" are currently legal. In these instances, trials would serve little purpose, except to shape public opinion.

I can think of a combination of measures that might, unlike trials, prevent the worst corporate abuse before it happens. (1) Return to the idea of corporate charters that have a fixed time period, say twenty years, after which they expire unless renewed (like broadcasting licenses), after a searching review of the corporation's activities. (2) Eliminate, probably by constitutional amendment, the special privileges that corporations have gradually gained in courts during the past 150 years. (3) Close the legal loopholes that enable senior executives to dissociate themselves from the misdeeds of the companies they control. (4) Make it much harder for corporations to evade punishment by jettisoning divisions, changing names, merging with other corporations, or otherwise altering their identities. (5) Restrict the ability of multinationals to use world trade regulations to nullify national environmental and human safety laws. (6) Protect communities by limiting the rights of multinationals to transfer factory operations and large blocks of capital from country to country or state to state without warning and without evaluation of local impact. (7) Change the laws and regulations that allow the largest corporations to avoid paying their fair share of the taxes needed to support the people and environment of the country. (8) Reflect on our own complicity in corporate violence; avoid purchasing products that we don't need and which are socially and environmentally damaging.

If human rights and a healthy environment are to make it through this twenty-first century, we had better do something soon to reestablish the authority of law over giant corporations. It will be an uphill battle, considering the trouble it took to shut down tiny Apple Tours. But the times are changing; a brighter light is shining on corporate boardrooms and down the corridors of power, even as the unwieldly size, inertia, and byzantine structure of the multinationals make them vulnerable to unexpected challenges from an angry public and a balky environment. It is too late, of course, for Randolph Walker and the numberless other victims of past corporate crimes. All we can do for them—and for posterity—is promise not to forget.

FOUR

RELATING TO NATURE IN A MANMADE WORLD

In relation to the earth, we have been autistic for centuries. Only now have we begun to listen with some attention and with a willingness to respond to the earth's demands that we cease our industrial assault, that we abandon our inner rage against the conditions of our earthly existence, that we renew our human participation in the grand liturgy of the universe.

Thomas Berry, *The Dream of the Earth*

THE VINE CLEANERS

First fragment. I am in my apartment, a modern, starkly decorated eyrie in a high-rise building. I know it is my apartment in the intuitive but unchallengeable way one knows things in dreams. Actually, I am in a short hallway within the apartment—a parqueted floor and a window at the end of the hall stick in my mind. There is a storm going on, but the other side of the window seems very far away. It grows dark outside; lightning flashes intermittently. Sam, my youngest child, warns me to stay away from the window; he points out that there is a danger of tornadoes. As if to underscore the seriousness of the threat, the lights of the apartment begin to flicker.

Second fragment. My wife, Joan, meets me at the door to the apartment. She looks upset.

"The vine cleaners are here," she whispers. It is clear to me that she wants to send them away—we don't need vine cleaners for our potted plants—but it is too late. Looking past Joan, I see that the vine cleaners, an immigrant couple, are already at work, spraying and polishing the leaves. I can smell the perfumed chemical cleanser they are using. Perhaps, I think to myself, this is a free service provided by the management of the apartment building; but no, it isn't, for now they are done and the man tells me that the vine cleaning costs $8.50. I hand him a $10 bill and get a one and two quarters in change. The couple waits. Do I tip them? A quarter each is hardly enough. Is a dollar too much? I hesitate. Outside, the storm rages. Then the lights flicker again and go out, freezing the tableau.

I opened my eyes with mixed feelings of relief and disorientation. Relief because the vine cleaners had vanished, along with the question of how much to tip them—there is no need to tip people in dreams. Disorientation because the place I had awakened in was so utterly different from the clinging aura of the dream, and yet this real place was also for a moment strange and unfamiliar. For I was not in my first-floor bedroom in New Jersey; I was, I gradually remembered, wrapped in a sleeping bag on a camp bunk in a one-room log cabin on an island in Lake Temagami, in central Ontario, Canada. The sun was shining and I could see the sparkle of the water through the birches outside the window.

It was the end of a five-day family canoeing trip, and we had spent the last night back at Northwaters, the wilderness canoeing camp whose gear and canoes we had used. We had started in mid-August, after the close of the regular camp season, so that our daughter Kate, a counselor at Northwaters, could lead us. Our son Jon, at that time twelve years old, had just finished a seven-week session at Langskib, the brother camp farther up the lake, and he came along too, eager for more even after several hundred miles of canoeing that summer. With Joan, Sam, and me, we made a party of five—Jane, our other daughter, was busy being a camp counselor for pre-kindergarten children in New Jersey and couldn't be with us.

Kate had just returned from co-leading one of Northwaters' wilderness expeditions, an environmental leadership training trip for ten young men and women down northern Manitoba's Seal River to Hudson Bay. Listening to her stories about forest fires raging along the banks upriver, polar bears prowling around the campsites farther down, and the early morning rendezvous in foggy Hudson Bay with the fishing boat that would take them to Churchill, I had had no qualms about going out on our thirty-mile lake paddle. But our trip also had had its moments, including the roaring wind and big waves on Lake Obobika.

Kate in action was a revelation, whether she was carrying a seventy-pound canoe with a thirty-pound Duluth pack on her back at a fast clip along a half-mile portage (she is five feet three inches tall, and slender), deciding when it was safe for us to be on the water and when it was not, reassuring nervous parents, baking cakes and bannocks in a reflector oven by the campfire, showing Sam how to care for himself and be help-

ful, and subtly guiding Jon's ongoing wilderness training without under-
mining his self-confidence. Jon was a revelation, too. A wonderful ca-
noeist, he carried huge loads on the portages and regaled us with an endless flow of jokes and stories. Suddenly mature and knowledgeable,
his seven weeks at Langskib had taught him that to be manly is to be
generous, resourceful, strong, fun-loving, and affectionate—traits also
shown by his very womanly sister.

During the trip, a curious reversal of roles had taken place. Joan and
I were accustomed to receiving good advice from Kate; she had been
giving it since she was three years old. But up to now we had always been
in charge. In the Canadian woods it was different. Here is an example.
It was our last portage; we were in the middle of moving our gear when
a large group of campers pushed past and piled up their canoes, partially
blocking the end of the narrow trail. I spoke angrily about it to one of the
campers, a boy of fifteen or sixteen, who reacted with embarrassment
and confusion. Then Kate took me aside.

"What were you trying to accomplish by talking to that boy?" she
asked me in a quiet, stern voice. I couldn't reply, partly because I was
ashamed of the answer, and partly because it was painful to be repri-
manded by my college-aged daughter. A little later, I heard her chatting
with the other group's counselor, out of earshot of the campers.

"Because you aren't from this area," Kate said, "you probably don't
know that it's a Lake Temagami custom to wait at a portage until the
previous group finishes." The counselor expressed surprise and thanked
Kate for the information, saying that it wouldn't happen again. And I
began to understand a little better that it takes more than canoeing skills
to make a good trip leader.

Lying on my bunk on this last morning of our short vacation, I
thought of the things we had seen and done on the trip. I thought of the
great, free-standing pillar of stone, sacred to the native peoples, that is
known as Conjuring Rock. I thought of the blueberries, last of the season, at our Lake Obobika campsite. I thought of the old-growth stand
of giant white pines with the small decorated parcels of birch bark and
twigs placed carefully against the trunks of the largest trees by unseen
hands, probably Ojibway. I thought of the big cat (I was sure it was a
lynx; everyone else said bobcat) that swam across the bows of our two canoes, then, gaining the shore, turned calmly to look at us before bound-

ing away. I thought of my fear turning magically into joy as I headed the canoe into Obobika's white-capped waves.

Only half-awake, I also thought about what David Knudsen, the founder of Langskib and Northwaters, was achieving: the use of wilderness to teach each child how to be fully human, a *mensch*, a person of honor, integrity, and compasssionate judgment. Nor was the teaching only for healthy kids. A year earlier, Kate had led an adult women's group on a highly successful trip. Other counselors guided trips for handicapped children, men's groups, business executives, even military personnel, but the lessons of cooperation, decency, and love and awareness of the environment were always the same.

But what of the vine cleaners? I couldn't get them out of my head. How did they fit in here, on the shores of Lake Temagami? It was my dream; nobody had imposed it on me. What was I telling myself? Was it an anticity message? I didn't think so; I like cities and know their value. The dream wasn't about cities; what was it about?

The vivid contrast between the life in the dream and the simple cabin where my family was sleeping held at least part of the answer. How can we remain human in a totally contrived world, where our days are cluttered with petty tasks whose sole function is to shore up and touch up the illusion we live in? I have seen the brilliant flowers of scarlet runner beans—quintessential vegetable crop of northern cities—lovingly tended by tenement dwellers in Manhattan's Bowery district, but I have also seen employees of a plant service company, on their monthly rounds to skyscraper lobbies and elegant apartments, replacing dying rubber trees in fancy flowerpots with fresh rented plants. My vine cleaners belonged to this latter group, touching up the illusion with their absurd spraying and polishing.

The fact that I made the vine cleaners an immigrant couple was not lost on me, either. When we fill our lives with what the late economist E. F. Schumacher called bad work, we invite exploitation and abuse of others, especially those who are most vulnerable. The awkwardness about the tip was another way of saying how hard it is to deal with people when we know that the work we have imposed on them is trivial and unnecessary. This applies not only to our employees but also to our friends and to our children. Bad work comes, in part, from being cut off from nature. In wilderness it is easy to see what matters in our lives and what doesn't. But we don't need to be in wilderness all the time to re-member what matters; with constant attention, contact with the natural world can be maintained wherever we dwell.

In the dream, a storm was raging outside the apartment; there were tornadoes. The lights were flickering, as if to show me the final depend-ence and fragility of my constructed environment. One of the conse-quences of our estrangement from nature is the belief that environmen-tal events need not touch our lives. I did not even have the sense (or strength?) to move away from the window; my nine-year-old child, Sam, had to urge me to step back. At the end of the dream, the lights in the building go out for good, leaving me, Joan, Sam, and the vine cleaners helpless in the dark.

We left Temagami that morning, heading back to suburban New Jer-sey. The taste of the dream lingered as I dressed and packed, as I said good-bye to the staff, and as I walked to the dock. It lingered as we waited for Jon and his counselor, Dave Torres, to give each other a fare-well hug. It lingered as the island slid away behind us. And it lingers still as I live and go about my daily business.

A CONNOISSEUR OF NATURE

My old college housemate Claude had dropped by for one of his stimu-lating visits, an infrequent happening even though his apartment on Manhattan's Upper West Side was only an hour away by car. In college, we had been utter opposites in interests and temperaments—Claude's direction was art criticism, which he celebrated with an élan that gained him a wide circle of friends. Mine was biology, and shyness was my dis-tinguishing social feature. A shared liking for history and the attraction of very different personalities formed a narrow bridge between us, enough to have kept us in occasional contact over the years.

We had been discussing European politics and the question of de-centralism over a light lunch on the patio, and then, in one of those cu-rious shifts that conversations take, we had abruptly changed the subject. He was holding forth in his usual compelling fashion.

"The trouble with nature writing," Claude said, "is that it's so artistically accepting, so uncritical. Lopez, Abbey, and their ilk—why civilization seems to have passed them by without leaving a mark. Culture implies the ability to discriminate, to judge—not sometimes, all of the time. Why should nature be exempt from criticism?"

He readjusted his long, elegantly trousered legs carefully on the lounge chair, glared briefly across the patio at a song sparrow sitting on a dead branch of my Cortland apple tree, and continued. "I fail to comprehend why you nature-lovers don't hold nature to the same kind of exacting standard that a cultured person applies to a bottle of Margaux or a performance of *Rosenkavalier*. Undiscriminating adulation is barbaric."

Just then a starling arrived at the bird feeder, displacing a blue jay. "Do you see that?" cried Claude. "That is exactly what I mean."

"Too aggressive?"

"No, no, that's not it at all. Look at the preposterous thing. Surely you can see what I'm referring to?"

"All I see is a starling. It seems healthy."

"It seems healthy!" Claude stared at me in disbelief. "You're not understanding me. Try to look at it as if it were a painting, a painting you did in beginner's art class. I am your instructor and we are examining it together. Now what's wrong with it?"

"I don't see anything wrong with it. It looks just like a starling."

Claude shook his head, closed his eyes for a moment, then cut himself a small wedge of brie from the round that was on the little table at his elbow.

"I apologize," he said after swallowing the morsel of cheese, "I was asking too much. Let's start with the head. What do you notice about the head?"

"It's a different starling," I observed, "the first one flew away."

"Never mind about the first one! I don't give a damn if it's the first one, the second one, or the fiftieth. Don't you notice that its head is all wrong? It's not the right size for the rest of its body. If Michelangelo's *David* had been proportioned like that, they would have cut it up for souvenir paperweights years ago. And that garish yellow beak—badly colored and crudely shaped. No decent design studio would tolerate that sort of work."

I saw what he was getting at, and nodded. Encouraged, he went on.

"About the absurd shape of the bird, there is no need to comment. But I must say something about the feathers. Those speckles. That cheap, iridescent shine—like a $79 suit in a sleazy men's discount store. Aesthetically, the bird doesn't work at all."

"They're very successful. There are millions of them in North America."

"Common," sniffed Claude. "Their numbers don't impress me."

I tried to be conciliatory. "You have to remember that they're an exotic species. They don't belong here. Somebody brought a few over from England and let them loose in Central Park in New York because they were mentioned by Shakespeare."

"That's quite irrelevant. True aesthetic sensibility is universal. It doesn't recognize national boundaries. I'm sure they are just as hideous in England."

"I don't think that Shakespeare ever said anything terribly negative about how starlings looked."

"Shakespeare had more important things on his mind: ungrateful daughters, jealous husbands, tyrannical usurpers, malicious servants. If he had had time, I have no doubt that he would have been severely critical of starlings. 'That ill-visag'd fowl.' Perhaps he might have worded it slightly differently."

I tried to come up with some ersatz Shakespeare to make the case for starlings—even though I don't much like them myself—but I couldn't compete with Claude. Still trying, I went inside to the kitchen to fetch glasses of iced tea and brought them back out to the patio. For a while we spoke about our mutual friends from college, what they were doing, how much they had aged. The song sparrow had been trilling loudly during this part of our conversation; Claude broke off his remarks and made a face.

"What a racket. That brings us right back to what I was saying earlier. If that were an instrumental piece, the critics would unanimously call attention to its total failure to explore musical space, its slavish commitment to melody, its inability to expand into another temporality. But because it's a bird, a privileged part of nature, it can commit a cultural felony and get off scot-free. And look at the grubby brute. What has it been doing to its chest? Has it been carrying a leaky fountain pen?"

"Every song sparrow is like that," I protested. "The dark spot on the breast is characteristic, and . . ."

"You don't seem to understand. Just because they *all* have it is not an excuse. But let's get off the subject of birds. I can see that you're looking at them through rose-tinted binoculars." He paused briefly to let me appreciate the joke. "I'm trying to make a general point. We have remade the world from top to bottom. Our canals cut from one ocean to the next. If there is something we want inside a mountain, we remove the mountain. If we need an offshore island where there isn't one, we make an offshore island. If a forest has been used up, we plant another in its place."

"Actually . . ." I interposed, but Claude held up his hand.

"Let me finish. The world is now largely a product of our own making. And like any artist or designer, we take the credit or blame for what we produce. If its form and texture are attractive and pleasing, we are deservedly praised; if it is ugly and tasteless, you can be sure that we will be harshly criticized. Why shouldn't nature be held accountable as well?"

"Actually, those planted forests don't work very well. I mean often they don't even grow, like the attempts to replant the cut-over areas of the Cascades in Oregon. In some places they tried and failed three times. And when the planted forests do grow—like the slash pine forests in Florida and Georgia—they're very monotonous and sterile."

"Then don't plant them. Didn't you once tell me that the English moorlands used to be forested? Think how many poets and novelists have written about their wild beauty. Obviously they're much better off the way the early Britons left them: without trees. They had excellent taste, the Celts and the Saxons, or whatever they were."

"Are you suggesting that the deforestation of the moorlands was a planned and deliberate creative act, like building Stonehenge?"

"Why not? We don't have any way of knowing what those folks thought. They may have been very artistically perceptive. If the forest was offensive, why shouldn't they have removed it for that reason and replaced it with something nicer?" Claude broke off his remarks, got up, stretched, and walked over to the patch of lawn that borders my vegetable garden. "Your little bit of moorland," he said, smiling.

I wasn't going to be baited. "Claude, you seem to believe that nature is just another collection of objects to be judged, like pictures in an exhibition. Do you really think that our aesthetic sense exists completely apart from nature? I've always understood that our idea of the beautiful is derived from natural forms in the first place. The Japanese provide examples. Their sand gardens, raked into patterns like ripples on a lake or wavy lines of surf on a beach. Their bonsai, imitating ancient, wind-pruned trees on a rugged mountainside . . ."

"Their Noh performances. Their tea ceremonies. Oh, come now. You're confusing nature nostalgia with aesthetics. I tell you humanity has expanded beyond nature. The most highly developed appreciation of the beautiful in art that I have ever encountered belongs to an urban friend of mine who once told me that he never left Manhattan until he was eighteen, when he went on a day hike with some college buddies in Harriman Park. He came home an hour and a half early because he couldn't stand it. But if I wanted a sensitive evaluation of a Japanese landscape painting I was thinking of buying, his opinion is the only one I would trust without question. Do you think he acquired this superb ability by looking at those nasty trees along the sidewalk on East 64th Street? Or maybe you think there was an influential potted rubber plant in his parents' apartment when he was a child?"

"Claude, haven't you ever enjoyed nature, been in a natural environment that you could savor, that made you feel good?" I asked. "I mean, made you feel good in a way quite different than you would if you took a trip to the Louvre?"

"Certainly I have," Claude answered. "I went to the Grand Canyon last year. It was quite nice. And some parts of it were really well composed, with a play of light, color, and line that was nearly the equal of what you might find in an early Daumier. Of course, Daumier didn't bother much with landscapes.

"Mind you," he continued, "the Grand Canyon is not beyond criticism. Its scale is too big to allow it to make any kind of modulated or nuclear artistic statement. That's a problem, but if you are willing to restrict your view while looking down from the rim, not try to take it all in like the silly tourists swiveling their heads from side to side, then it's possible to have a pleasant experience."

"Was that the only place you went on your trip?"

"No. I traveled about other parts of Arizona—my friends insisted on it. I saw the giant meteor crater and then made my way south."

"What did you think of the meteor crater?"

"Well, you know it wasn't nearly as attractive as that open-pit copper mine near Bisbee. They're both craters, I grant you, big round holes in the ground, but I found the lines of the Bisbee mine, with those striking terraces around the sides, much more satisfying. That's a case where nature might have done better by imitating art."

I was struck by the thought that a copper mine, which I view as an environmental blight, was for Claude a thing of beauty. Clearly, natural context was of no importance to him. How could I explain this? I decided not to try.

For the last few minutes, Claude had been mopping his face with a paper napkin. "I think we ought to go inside," he said; "it will be cooler there."

"Only a little cooler. The electric company has imposed a brownout. Too many people using electricity during this Labor Day heat wave. I'm afraid to turn on my air conditioner—it might be damaged by the low voltage."

"I don't see why they don't build more power stations if they can't supply enough power," Claude responded irritably. "It's a perfectly simple engineering problem. Well, I think I'll drive back to the city. The air conditioner in my car is working, at any rate. Many thanks for your hospitality; I enjoy our visits."

"I'm sorry about the air conditioner, but nature does impose some limits," I said. "We can't just change the whole climate of the earth if we're too hot."

Claude laughed and patted me on the shoulder. "There you go about nature again. You really are an incurable romantic. We made the world hot with global warming, as you yourself pointed out to me the last time we got together. Now we'll just have to cool it off again. We'll find a way. Bye."

There was no chance for me to respond. Claude was already striding towards his car, which was parked at the far end of the block. His earphones were in place, and the cassette player fastened to his belt was switched on. I turned away. Then, as I walked back to the house musing about our conversation, I suddenly realized just how Shakespeare would have phrased a complimentary remark about starlings, if he had had the time. I would tell it to Claude during his next visit.

DEATH OF A PLASTIC PALM

All over the United States, phone companies are planting artificial trees. An old-growth, imitation pine has just gone up in the community of Franklin Lakes, New Jersey, where it towers 100 feet over barns and fields. Holes in its synthetic brown bark admit bundles of black cables that travel up through the hollow metal trunk to the wireless digital phone antennas hidden among the plastic pine branches high above. In the desert near Phoenix, Arizona, where pines are scarce, another phone company plans to conceal its antennas inside a giant, imitation saguaro cactus. Evidently some large corporations are finally beginning to realize that the destruction of nature by their business activities is not acceptable. To improve public relations, they are now providing facsimiles of the nature they eliminate. This solves many annoying problems, but, alas, other difficulties have arisen. Copying nature is not as easy as it looks—it's not like using a photocopier. If the phone companies don't know this, somebody ought to tell them.

I first became aware of the downside of artificial nature when I was living in Middlesex, New Jersey. To get to work at the university I had to drive through South Bound Brook, a down-at-the-heels little town nestled cozily between American Cyanamid's huge organic chemical factory on one side and GAF and Union Carbide—also cooking up organics—on the other. True, the Raritan River runs alongside of South Bound Brook, and George Washington, an experienced surveyor who had seen many rivers, once described the Raritan as the most beautiful of all. But the river's magic seemed to work feebly in South Bound Brook; my memories are only of floods, brown mud, and the occasional corpse of carp or sucker floating belly up past the dirty red brick of the aged factory walls.

Then one day as I made my daily commute, I noticed an unexpected splash of color in the town. A popular saloon along the road I traveled had erected a very large plastic palm tree on the sidewalk next to the entrance. Its ornate fronds swayed gently in the torpid breeze, beckoning to the factory workers passing by; if the green of the leaves was a shade more vivid than one generally finds in nature, and if the bark was a trifle more orange, nevertheless the palm was clearly making an important statement. From then on, every morning on the way to work and every evening coming home I looked at that palm, with feelings one could liken to those of John Bartram and his artistic son Billy as they gazed upon their great discovery, the elegant, flowering *Franklinia*, growing alongside the Altamaha River in southern Georgia back in 1765. The plastic palm became a fixture for me, a necessary part of the landscape, standing in for the shaggy-barked, curving river birch or massive, buttressed sycamore or pin oak that I might have seen in South Bound Brook had I made my trips in the time of the Bartrams.

Months passed, perhaps a year and a half, before I noticed one day that something was wrong with the plastic palm. The leaves were yellowing. The change was slight, observable only at the tips of the fronds, but to one who is keenly aware of nature, there could be no doubt. Although I am not an expert on plastic, I do know something about real plants. When the leaves of an evergreen tree like a pine or palm start to lose their chlorophyll it is a sure sign of illness—ask any plant pathologist. During the next few weeks I cheered myself with the hope that the palm was only suffering from a transient nutrient imbalance. Maybe it will recover, I thought. Give it some time.

But its condition only grew worse. The yellowing spread, covering all the leaves; soon the friendly, garish green was just a memory. Then the broad leaf blades began to disintegrate and fall away, each leaving behind its central, supporting rachis. By autumn, not a single frond was left intact. The once-leafy crown of the palm was now a forlorn collection of bare plastic (or aluminum?) wires drooping out of the top of the trunk. It reminded me of a coconut palm that used to grow on the beach in Tortuguero, Costa Rica, in front of our research station. On an August morning in the mid-1960s, a group of three or four CIA-supported anti-Castro freedom fighters, who used our little airstrip on the beach to fly in toilet paper and other supplies for their camp upriver, shot away most of the leaves of the coconut with machine guns. We never knew why they did it—perhaps the tree had offended them in some way—but whatever their motives, when they were finished it looked a lot like the plastic palm in South Bound Brook during its last few months.

I could no longer bear to see the afflicted palm and averted my head when driving past, traffic permitting. At least a year went by until the saloon owner had the remains removed. Just before it went, I noticed out of the corner of my eye that the bark was splitting and peeling off in tattered strips.

Eventually I forgot about the plastic palm, but was reminded of it in 1997 when I saw another one in front of a store on Route 36, on the way to the beach at Sandy Hook. I make that trip often, taking students on field trips, and each time I have looked carefully at Palm II, as I refer to it. Sure enough, on the last few trips I have noticed the now unmistakable signs of yellowing at the tips of the fronds. It's progressing slowly, but I know what the outcome will be. Palm II is dying. This time, instead of grief I felt a thrill of excitement: the death of the South Bound Brook palm had not been a special case. I am the first scientist to discover a disease of imitation plants.

What is causing this apparent epidemic among plastic palms? Not knowing where else to turn, I consulted the fountainhead of information about tree diseases, *Tree Maintenance* by P. P. Pirone, 4th ed. Going directly to the section on palms, I soon came across *Pythium* root rot, which "is accompanied by yellowing and wilting of the leaves one after another, until the bud falls from the top of the plant." Admittedly, there is some similarity here, but I don't really think that the plastic palm wilt (this is what I have named my new disease) is fungal, like *Pythium*. Air pollution, which has been known to affect and kill plants for 150 years, is a more plausible culprit. Sulfur dioxide, fluorides, hydrocarbons, ozone, peroxyacetyl nitrates, nitrogen oxides, ammonia, and assorted acids are all toxic to plants, and all are common enough in urban and even in some rural air. A glance at Richard Harris's important book, *Arboriculture*, confirmed my suspicions: "Leaves are the plant parts most likely to show symptoms of air pollution injury," says Harris, and I have no reason to doubt his word. Perhaps the plastic palm wilt will turn out to be caused by air pollution.

The implications of my discovery are disturbing. After all, why take such pains to imitate nature if we are going to have this kind of trouble with its replacements? What went wrong, I wondered? Surely the original idea of reproducing only the brightest and most beautiful inventions of nature—green leaves, waving branches, stately tree trunks—was a sound one. There must be a catch, I realized. Could it be that nature packages its inventions? Do the parts belong to an elaborate, interlinked system that can't be completely disassembled? Of course, that's it. Pull out the green leaves and waving branches for copying, and like it or not along come some other pieces of nature: rot, decay, even a pretty good facsimile of mortality. Wow! I thought, is the phone company aware of this? If there can be a plastic palm wilt, why not a plastic pine blight or a steel cactus canker?

Then an even more worrisome idea occurred to me. When we copy nature, not only do we get things we don't want, but some critical parts are always missing, parts that we can't copy, the best parts of all. These are the signature processes of nature: self-repair, regeneration, and re-birth. Nature holds on to them tightly; they are not just copyrighted, they are copyproof. Decay and death are tolerable, even beautiful, in nature because they are food for regeneration and rebirth. In imitation nature, decay and death lead nowhere—uncoupled from the grand cycle, they are shabby, pointless, and unredeemably grim.

It almost seems as if nature does not like synthetic imitations. Imitation may be the sincerest form of flattery, but nature is neither flattered nor amused. Her anger is a serious matter. That's why I am concerned about the 100-foot plastic and metal pine in Franklin Lakes. The machine-gunned coconut in Tortuguero eventually put out new leaves, but what will happen when the imitation pine loses its branches in a high wind or ice storm, as it inevitably must? Will new branches sprout from the trunk; will the broken giant be ringed by imitation pine saplings reaching upward in the sudden sunlight to replace it? No, the phone company will have to pay to have their pine fixed. And pay and pay and pay and pay forever for what once was free.

SCIENTIFIC DISCOVERIES AND
NATURE'S MYSTERIES

One of the joys of science is its steady production of wonderful discoveries. At irregular but not intolerable intervals, it throws out to the waiting world findings that add spice to a thousand morning newspapers and enliven the conversations of countless barbers and taxi drivers. Some branches of science are especially fruitful in their production of wonders: physics doles out an inexhaustible supply of elemental particles with clever, memorable names like *quark*. Medicine provides powerful antibiotics freshly isolated from weird sources such as exotic muds and frogskin, arraying them against an equally fresh crop of antibiotic-resistant microbes. Geology reveals previously unsuspected earthquake faults under major cities and dazzles us with hidden, underwater chains of volcanoes in secret, violent eruption. Even paleontology can be counted on for the latest, well-documented theory of how birds came to fly or, as I have already discussed, why the dinosaurs became extinct. And then, every ten years or so, like the return of a speeded-up Halley's Comet, animal behavior brightens the scientific horizon with a new and incredible explanation of how animals find their way home.

Not long ago, the wires were humming, or would have hummed if wires hummed anymore, with the discovery by P. Berthold and three colleagues that bird migration has a rapidly evolving genetic basis. Evidently, the inherited tendency of a population to choose a particular migratory direction can change in a few years, with the selection of another direction. Berthold, working in Radolfzell, in southwest Germany, made the most of an earlier observation: since the 1950s, increasing numbers of blackcaps, Old World warblers, have been spending the winter in Britain, traveling from their breeding grounds in Germany and Austria during the fall migration. Traditional wintering grounds are in the western Mediterranean countries far to the south; most blackcaps still go there in the fall. But the blackcaps wintering in Britain have done well; global warming (or chance) has ameliorated the winter climate, and the British are dedicated bird feeders.

What Berthold did was to capture forty male and female blackcaps in England, transport them to Germany, breed them, and raise their young. A group of nestling blackcaps from nests around Radolfzell, presumed to be mostly or entirely from the Mediterranean wintering population, served as a control. In the fall, when the young birds of both groups were fully fledged, they began to show migratory restlessness and to point in the direction they wanted to fly, a phenomenon noted many years earlier by the ornithologist Gustav Kramer. Berthold put them separately in special, outdoor, circular cages that recorded which way the birds were facing. The British-bred birds pointed slightly north of west, toward Britain; the Radolfzell controls faced southwest, toward Spain.

There are a few flaws in the experiment, as there are in all experiments: for example, it would have been better to supplement the Radolfzell controls with a group of captive-bred nestlings whose parents came from Spain. But no need to quibble. This experiment reinforces the older demonstration that the direction birds face when they want to migrate, probably the same as the actual departure heading, is inherited. More exciting, Berthold has shown that natural selection can quickly modify the inheritance of departure heading in migratory birds. In less than forty years, the proportion of German-born blackcaps that overwinter in Britain has gone from zero to around 10 percent.

Has Berthold's experiment finally answered the ancient question of how animals find their way home during long journeys? Hardly. Like the pot of gold at the end of the rainbow, the definitive and final answers are always over the top of the next hill. Looking back at a succession of great discoveries about animal navigation during the past half-century, I see a clear pattern, a regular sequence of events. It goes like this: discovery, scientific enthusiasm, public awareness and enthusiasm, supportive findings, complications, disillusionment. Notice that I use the word *complications*, not *disproof*. Major discoveries concerning animal navigation (and other natural phenomena) rarely get disproved, just complicated.

I should describe what I mean by navigation. This is the ability, shared by many species of birds, sea turtles, certain butterflies, various fish, and a host of other species, possibly including dogs, cats, and people, to find a particular place that is far away, while traveling over unfamiliar territory. True navigation is more than orientation to known and fixed landmarks. To a green turtle swimming in the South Atlantic, out of sight of land and far above the ocean floor, heading for Ascension Island, a five-mile speck of land a thousand miles ahead, there are no fixed landmarks—at least, not evident ones.

Nor does a mere "compass sense" suffice for navigation. A compass sense, being able to tell north from south from east from west, is probably shared by most species of animals, including invertebrates. It is a necessary part of navigation systems, but usually it can't do the job by itself. You can use a compass to find home only if you know what direction home is supposed to be, which requires the equivalent of a map and knowledge of where you are on that map, the "X—you are here" of hiking trail and shopping mall signs. I can describe another experiment that should make this clear.

More than a decade before Berthold did his work, a Dutch scientist, A. C. Perdeck, performed an even more spectacular research feat, with remarkable results. Each year in the fall, starlings that breed in the spring and summer in the eastern Baltic countries fly southwest in large flocks, traveling across the Netherlands on their way to their winter territory in southern England and Ireland and northern France. They fly in mixed flocks, with the newly fledged, inexperienced migrants accompanying adult birds who have made the trip before. Perdeck captured 11,000 of these migrating starlings near the Hague and carried them to Switzerland, far to the south of their familiar migratory route. There he banded them and released them, but he took care to release them in separate flocks composed of only adult or only immature birds.

Recoveries of banded birds revealed a striking difference: the immature, first-time migrants simply continued on the typical southwestern fall compass heading (as if they were still in the Netherlands) and ended up in southern France and Spain where they spent the winter, hundreds of miles south of their traditional winter habitat. But the adults flew in an unusual northwestern direction over unfamiliar territory, landing in their regular wintering grounds in northern France and, in a few cases, southern England. The adults showed true navigational ability; the juveniles merely had one component of that ability, compass sense.

The incredible conclusion to this experiment was that the following spring some of the younger but now mature birds returned to their place of birth in the eastern Baltic countries to nest. And the next fall, some of these birds migrated again—not to England, Ireland, and northern France with all the other starlings—but to southern France and Spain, where they mistakenly had spent their first winter.

If I have explained it right and if you have followed everything so far, you may have noticed that the results of the Perdeck experiment complicate the interpretation of the results of Berthold's later study. Perdeck showed that although an innate compass heading (including, by implication, the initial compass heading of migratory restlessness) is a factor in bird navigation, it is not really necessary; the adult starlings in his study abandoned it altogether when they were displaced off course to Switzerland, an unfamiliar location. The inborn compass heading was followed only if the adult birds *already knew* they were flying in the right direction.

The history of research in animal navigation has many other incredible discoveries and theories, always followed (or preceded) by equally incredible complications. I will quickly describe some more examples.

Birds and honeybees have an accurate sun compass based on visual detection of the position of the sun, plus an accurate biological clock to correct for the change in the sun's position with time. But the sun does not have to be visible for the compass to work; any patch of blue sky has a polarization pattern that gives the sun's location, as Karl von Frisch showed for honeybees. Bees, and probably birds, can see this polarization pattern in the sky; people can too, if they work at it for a few years. Yet if the polarization pattern is obscured by total overcast, bees and birds *still* have an accurate compass.

Some species of birds can navigate at night. Franz Sauer, one of my teachers, first put them in a planetarium and showed that by rotating the night sky he could make caged warblers change their heading in a predictable way. S. T. Emlen extended this work to identify the important features of the night sky for indigo buntings; they paid particular attention to the area around Polaris (the North Star) and the Milky Way. Yet nocturnally migrating birds can hold a true heading while flying in thick clouds on an overcast night when no stars are visible, as radar studies have shown.

Floriano Papi and his colleagues demonstrated that homing pigeons can use their sense of smell to find their way to the home loft when it is a few miles away; he came up with a simple but brilliant theory to explain how this olfactory guidance system could work, even if the wind is blowing from the wrong direction. Yet pigeons can find the home loft when it is hundreds of miles away, out of range of familiar smells. Moreover, Papi's experiments, which worked well near Florence, Italy, did not work in Ithaca, New York.

The earth's magnetic field can be detected by compass needles, snails, honey bees, sea turtles, birds, and many other animals. Excitement mounted when James Gould and Charles Walcott found crystals of magnetite, magnetic iron oxide, in the heads of homing pigeons. Yet it is extremely difficult to get bicoordinate map information (latitude and longitude) by measuring magnetic fields; the many experiments done to test magnetic homing in pigeons and people give circumstantial and conflicting results. There is still no good explanation of how an animal's sensory and nervous system could accurately measure the horizontal and vertical components of magnetic fields, let alone turn the incredibly complex data set into a map.

My favorite explanation of animal navigation is inertial guidance, the navigational system of rockets, which received much attention from behavioral scientists around the time of the Apollo moon voyages. Inertial guidance (or "dead reckoning") requires both a map and compass sense; an animal would have to register, remember, and analyze the duration, speed, and direction of every phase of its travels, comparing its position against its "map." This was first suggested in a few paragraphs of a letter to *Nature* dated April 3, 1873. The author, Charles Darwin, wrote,

> We must bear in mind that neither a compass, nor the north star, nor any such sign, suffices to guide a man to a particular spot through an intricate country, or through hummocky ice, when many deviations from a straight course are inevitable, unless the deviations are allowed for, or a sort of 'dead reckoning' is kept. All men are able to do this to a greater or less degree, and the natives of Siberia apparently to a wonderful extent, though probably in an unconscious manner. . . . Whether animals may not posess the faculty of keeping a dead reckoning of their course in a much more perfect degree than can man . . . I will not attempt to discuss, as I have not sufficient data.

The great naturalist Henry Walter Bates and a companion had their lives saved by a ten-year-old Indian boy who was traveling with them, when they wandered away from the Amazon River, where they had moored their canoe, and became lost in an unfamiliar part of the rain forest. The boy, who had been busy playing with a bow and arrow during their increasingly frantic wanderings, guided them directly back to their canoe when Bates finally thought to ask him for help. "I believe he had noted the course we had taken almost unconsciously," wrote Bates after the adventure. Yet it is hard to imagine such a system of inertial guidance, if that's what it is, working for an arctic tern during its yearly round-trip migration of 25,000 miles, one-tenth the distance to the moon.

Finally, there were the studies of the Canadian fisheries scientist, G. Groot, who was trying to understand how juvenile sockeye salmon find their way through a complex maze of interconnected lakes to reach the Pacific Ocean. Having shown that they were perfectly capable of orienting themselves using celestial cues, including polarized light, under clear or partly cloudy skies, Groot demonstrated that they did not need these cues to orientate properly under completely cloudy skies or in the laboratory. Perhaps, he thought, inertial guidance was involved—the salmon smolts had memorized all the twists and turns and changes in speed on the way to the enclosed experimental tank. But when he transported them under deep general anesthesia they oriented as well as they did back in their lakes under a clear sky. In the end, he described their orientation system as "type-X orientation," and left it at that.

And we are left, after a heady succession of discoveries, theories, and complications, with the gleaming, undented core of a great mystery. Is it, as some have suggested, that orientation mechanisms already described, working in tandem, often redundantly, in various combinations, can explain all feats of animal navigation? I doubt it. Are there fundamental phenomena yet to be discovered? Perhaps. Only one thing is sure: the methods of reductionist science skillfully applied over many years have not solved the problem of how animals find their way home. Is it permissible to wonder whether the reductionist approach *can* answer the important questions life poses? At any rate, I look forward to more years of discovery and disillusionment in this exhilarating, ever-daunting quest for a deep understanding of the mysteries of nature.

I REINVENT AGRICULTURE

Many of us who work with pencil, pen, or computer like to vary the intellectual routine by doing something "real." Some build cabinets and canoes, some sing in choral groups, others join first aid squads. I take care of our garden. Gardening is an especially good activity for eggheads because it combines physical labor with the opportunity to be creative. Unlike the ideas in the books I write, however, my creations in the garden are quickly put to the test. My experience last summer is a case in point.

I like to eat what I grow, so I have little patience with flowers and other ornamentals. I tolerate the ones that were planted around the house when we first arrived: yew, hemlock, cherry laurel, forsythia, flame azalea—the usual things provided by botanically challenged, cost-conscious builders. If these woody perennials get sick I do what I can for them, given my lack of knowledge and low level of interest. Sometimes I fertilize, sometimes I continue to refrain from fertilizing, depending on my intuition in the particular case. The same goes for watering. Our flame azalea is down to one tiny branch with eleven leaves on it, about the same as last year. I don't know whether I am keeping it alive against all odds, or preventing it from making the full recovery that it would be capable of if cared for by somebody else. I view my role in the treatment of our ornamental plantings like that of the remote HMO clerk who makes the critical decisions in a complex medical case; any unfortunate consequences of my ignorance are not of great concern.

Fruits and vegetables are another matter. I am not much more knowledgeable about them than about azaleas, but I try a lot harder. Because this is a story about vegetables, I will skip the fruits, except to note that we pick our own sour cherries, figs, peaches, plums, grapes, apricots, and apples—in sharply decreasing order of abundance (squirrels got all of my apples this year). Enjoyable as it is, however, to eat my breakfast cereal with figs from the garden (in New Jersey!) in October, it is the vegetables that provide the deepest satisfaction. Not that I like collards and potatoes more than fresh figs and iced sour cherry soup—I don't—but it is in the vegetable garden that I have been able to make my contribution to world agriculture.

Perhaps I should rephrase that. I haven't discovered anything that millions of gardeners and farmers didn't already know. It is not a contribution in that sense. But I discovered it for myself—I didn't read about it in a book or hear about it from another gardener. What I discovered is the secret of feral vegetables. (Some people call them "volunteers." You will soon see why this is much too benign a label.)

Feral, in the way that I am using the word, means having escaped from domestication and become wild. One hears about packs of feral dogs, ownerless, wary of people, living off the land in suburban and rural woods, and occasionally attacking children. Or solitary feral cats indulging their natural love of independence and wreaking havoc among local rodent and bird populations. My first feral vegetable was a patch of coriander, also called cilantro. Originally, I had planted coriander seed purchased from a seed house in Oregon. It grew fairly well and we ate it in guacamole and curries. But we didn't eat all of it; a few plants flowered and produced seeds that fell on the ground and lay hidden in the soil. The next year, I planted more coriander in a different part of the garden. Several weeks before it came up, I noticed scores of little corianders shooting up near what had been last year's patch and had now become the space between two rows of potatoes. The feral coriander was much more vigorous and luxuriant than the planted stuff: so abundant that it supplied all our requirements for the herb, with enough left over to flower and go to seed again. This happened years ago, and there has been no need to plant coriander again. Although I am never sure exactly where it is going to appear, there is always plenty of it somewhere in the garden. For a deeply lazy person like me, this is an ideal arrangement.

Soon I discovered other annual vegetables that were happy to escape domestication and lead a feral existence. Oriental red mustard greens were the first to follow suit; the feral plants, like the coriander, did much better than the ones I had planted. This, as I found out from repeated experience, is a general rule: things that plant themselves sprout earlier, grow faster, and have less trouble with pests than their planted and managed domestic counterparts. All that is required from me is the ability to recognize the feral vegetables at an early stage and avoid weeding them out, unless they are in the way of something I want more. Gradually, in my mind, a revolutionary new system of agriculture was taking shape; when perfected, all I would have to do is sit on the patio and let the feral vegetables raise themselves.

After the mustard came cherry tomatoes. If I planted their parents in a row next to my snow pea trellis, the following year little tomato plants would appear under the snow peas and be big enough to tie to the trellis in late June, after the snow peas were finished. Then beans joined the list of feral vegetables, plus an occasional potato, and even a pair of hefty collards, which didn't come from seed but from a couple of plants that had been overlooked when the row was cleared the previous fall. The garden started to look terribly disorganized, with miscellaneous clumps of vegetables coming up in unpredictable places; but the feral vegetables were feeding us, and that's what counted. Of course it was too good to last.

Early this summer, a new feral vegetable appeared in the space between two rows of tomatoes. It looked like a little squash plant, which puzzled me because I hadn't had squashes in that spot last year. Then I remembered that we had put down compost in the fall, and it might have included seeds from a supermarket Japanese winter squash of the buttercup type, our favorite. "Good," I thought, and left it alone. The squash grew rapidly, and I trained the vines out towards the edge of the lawn so that they wouldn't bother the tomatoes. Sure enough, when the first squashes appeared, they had the typical buttercup turban shape; eventually we had three or four tasty meals from them.

Meanwhile, other feral squashes were growing from the same clump. The leaves were different, much larger and silvery green in color; there were sharp prickles on the leaves and stems. They were rowdy vegetables: in no time at all they had leaped over the Rose, Aunt Tettrazine's, and Tommy Toes tomato plants and were heading south towards the Kennebec potatoes. "Good. More squash," I said to myself. I gingerly picked off a few of the gigantic leaves that were shading the tomatoes and left, with my wife, Joan, on a two-week trip to Glacier Park, Hell's Canyon, and Spokane.

When we got back in mid-August, the tomatoes and potatoes were gone. At least I couldn't see them under a billowing sea of squash leaves. To get into the garden I had to step on a meshwork of squash stems, which didn't bother them because each had the tensile strength of one of the cables holding up the George Washington Bridge, and about the same thickness. Then I saw it—a monstrous squash. It was long and yellow-green in color, with dark green stripes, a bit like a zucchini with elephantiasis. Looking around, I noticed others resting heavily on the potato hills or peering out from the mixed vegetation where I assumed the tomatoes still were.

What variety of squash was it, I wondered? Protected by cotton gloves, I cut a length of stem containing both leaves and a clublike fruit, threw it in the back of the station wagon, and drove to Oved Shifriss's house. I have mentioned Professor Shifriss before: long-retired from a distinguished career as a plant breeder, he still knows everything there is to know about squashes. He was resting when I arrived and seemed taken aback when I dumped my squash cutting on the top of one of his filing cabinets.

"Have you ever seen anything like this?" I asked.

"No," he replied in a tone that told me he would be happy if he never saw it again.

"What is it?"

"All I can say is that it's in the genus *Cucurbita*, and probably has genes from two very different species."

"Can I eat it?"

"I wouldn't," said Oved, politely indicating that I should take my squash with me when I left.

I figured out what had happened. The Japanese squash whose seeds we had put in the compost must have been a hybrid. Most of the off-spring of hybrid plants do not have the appearance and favorable characteristics of their parents. The reason for this has to do with chromosomes. Generally, chromosomes come in pairs (or multiples of pairs), and in a long-established variety, each chromosome in a pair is genetically much like the other. When two varieties are crossed to form a hybrid, the offspring receives one chromosome from each of the many pairs belonging to each parent. Therefore the hybrid predictably gets half its chromosomes from one parent and half from the other; and all the hybrids from that cross are similar to each other, although not to their parents. But when two hybrids are crossed—which is what happened with the seeds from my hybrid squash—a vast number of combinations of chromosomes are possible. Some second-generation hybrids, like the buttercup squashes we ate, are similar to their hybrid parents; most are not.

This means that if you want to grow the same hybrid vegetable again next season, you shouldn't save seed from your crop. Hybrid vegetables are favored by the big modern seed companies, because you have to buy new hybrid seeds every year. Agribusiness loves dependent gardeners and farmers; they are a perpetual source of cash. Of course, some hybrids offer value for the money, but I don't plant them if there is a good alternative. This is why my other feral vegetables look like their nonhybrid parents. But my lenience toward feral vegetables should never have been extended to seeds from the compost pile, where second-generation hybrids were waiting to sneak into my garden through the back door.

Grimly, I went out to the garden armed with a two-foot Collins machete. The feral squash leered at me, like the carnivorous plant in the movie *The Little Shop of Horrors*, or perhaps more like the intelligent walking plants that conquered the world with their lethal stings in John Wyndham's terrifying novel, *The Day of the Triffids*. It took a half-hour of hard work before the feral vegetable was lying on our patio—fourteen squashes totaling more than sixty pounds in weight, and at least 100 feet of vines. The stems and leaves were saved to put back on the garden as mulch, but the squashes, with their deceitful seeds, went out with the trash.

The potatoes in the garden were all right; they had been dying back in late July anyway. The tomato plants had fared worse. With the squash covering gone, they looked bedraggled and resentful. The Aunt T's still had tomatoes on them, but the other varieties were pretty much of a loss. It's just not safe to be an innovative gardener in these days of agricultural multinationals, I mused. But I can't blame everything on agribusiness. Deep down, I know that the giant inedible squashes were sent as a punishment for my hubris in thinking that I could reinvent in a decade a system of agriculture that my ancestors took nine or ten millennia to design. Next year, I'll limit my feral vegetables to the ones I can trust.

THINKING ABOUT BREEDS AND SPECIES

Our family dog is an Andalusian bog hound, a perfect specimen of his breed. His appearance and behavior are typical of bog hounds, but for those readers not familiar with them, a description is in order. An unusually handsome animal, he stands 22 inches high at the shoulder, maintains a trim weight of fifty-seven pounds, and has a short, smooth coat of a rich, red-brown color and a well-proportioned head with strong jaw musculature and brown eyes. He is extremely fast, and while running in open areas makes frequent, stiff-legged bounds straight up in the air. This may be an ancient canine technique of cooperative pack hunting, probably serving both to stir up prey and to provide a view over tall grass. His name is J. D., given by my daughter Kate because the way he moves his hindquarters when he walks reminded her of a character of that name played by Brad Pitt, in the movie *Thelma and Louise.*

J. D. is exceptionally affectionate. He has never been known to growl or snap at anyone, although he once took exception to a recording of Paul Robeson singing "Old Man River." When permitted, he greets people by standing on his hind legs with his feet on their chest (or shoulders, if they are short) and gazing directly into their eyes. He rarely shows affection by licking; as a special mark of favor he may take a hand or arm into his mouth and chew gently.

His distaste for swimming—as opposed to walking in muck—is characteristic of bog hounds, especially Andalusians. We tested this behavior during a trip along the headwaters of the Passaic River, in northwest New Jersey. Coming to a deep pool, my wife, Joan, J. D.'s favorite, jumped in and pretended to be drowning. J. D. watched from the bank in great agitation as she splashed and screamed, but he made it quite plain that if anyone were to jump in and save her, it would have to be me, or perhaps another dog.

It is hard to judge J. D.'s intelligence. The famous bog hound "seven-second delay" keeps him from responding promptly to difficult commands such as "sit down," but does not seem to slow his reactions to squirrels or food. Some experts believe that the delay is an evolved adaptation to travel in bogs, where a hasty misstep can plunge an animal into treacherous quicksand.

Apart from this slight peculiarity—not really a handicap—J. D. is an ideal dog, gentle with children and friendly to adults, yet an excellent watchdog with a deep, terrifying bark; not picky about his food; not at all neurotic; equally happy on a ten-mile hike or lying on a sofa. As the vet, on one of J. D.'s rare visits, described him, he is "a great, low-maintenance dog."

After this description, some of you may be wondering where you can get an Andalusian bog hound. But I'm afraid you can't. Kate picked J. D. up six and a half years ago while driving in the mountains of New Mexico, at a house with a sign in front that said "Puppies." (It was time, she told us later by phone from somewhere in Kentucky or Tennessee, that her brothers Sam and Jon, then seven and ten, had a dog.) The Navajo family that was giving away the pups hadn't been sure of their parentage, except that the mother was small. The puppies had purple spots on their tongues, indicating that the unknown father was a chow, or something like a chow. There might be some wolf in the puppies, they said, but this was probably romantic invention to make them seem more desirable than ordinary dogs. Since then, we have been told, by people who were fully satisfied with their own expertise, that J. D. is, besides chow, partly Akita, partly Alsatian, and partly several other breeds whose names I have carelessly forgotten.

Given the genetic uncertainty that obscures his background, why, therefore, do I call him an Andalusian bog hound? Well, it seemed the right thing to do. You see, J. D. is so unusually good-looking that people stop me on the street and ask, "What kind of dog is that?" I used to answer, "A mutt," but gave it up because I saw their obvious disappointment. I felt obliged to supply the name of a breed. But what breed? I couldn't call him a flat-coated retriever or an Australian sheepdog because sooner or later I was bound to run into someone who actually knew what these sorts of dogs look like. Much better to find a breed that no one has ever seen; after considerable thought, I invented the Andalusian bog hound. It has been gratifying to note how much this innocent deception pleases most questioners. "Andalusian" conveys the idea of foreign nobility, while "bog hound" turns J. D. into a specialist, a distinction to which most Americans accord instant respect. True, a few cynics remain who are suspicious of J. D.'s synthetic pedigree. To these I respond by hinting at his wolf ancestry; that helps to divert them.

This state of affairs continued for several years, and there would have been no need to write this account were it not for a discussion I had one Saturday in May with an acquaintance of mine. Paul also walks a dog, and people sometimes ask what breed it is. "That question makes me furious," he said. "What difference does it make what breed it is? If they asked 'Is he friendly?' or 'Does he bite?' I could understand it. Those are questions about what my dog is really like. But asking about his breed is like asking about somebody's ethnic background. Who cares what his parents were? This kind of racial stereotyping is what's causing all the murder, torture, and destruction in the Balkans and in Africa."

I was taken aback. I had never considered that promoting J. D. as a purebred dog might be a racist act. Then I remembered another incident that had raised the issue of stereotyping in a slightly different way. It had occurred the month before, while I was lecturing at Yale. During the question and answer period after the talk, the subject of the introduction of exotic species came up. I mentioned the damage caused by non-native species such as water hyacinths, Asian tiger mosquitoes, Argentine (fire) ants, and chestnut blight. Up went a hand in the audience. Was I aware, the fellow asked, of recent papers pointing out that the fear of alien species is no different from the racist hatred of newly arrived immigrant peoples? After all, he said, species have always moved from place to place. It's a natural process. Why single out the recent arrivals for condemnation? I tried to explain that introduced species, which find few predators, parasites, or diseases to keep them in check in their new environment, sometimes run amok. Water hyacinths clog waterways, tiger mosquitoes can spread disease, and so on. I don't think he was convinced.

Paul was outraged by the idea of stereotyping dogs by breed, and the fellow at Yale objected strongly to classifying certain plants and animals as exotic species. I asked myself, is there something wrong with using the categories of breed, variety, and species when we talk about living organisms? People love to classify things; it seems an innate part of human nature and often helps us make sense of our world. But perhaps it is a habit we should try to overcome in this new and hopeful age, when Global and One, rather than Local and Many, is the slogan, and genes can be moved, irrespective of the species differences, from cows to yeast.

Let me give this new approach a try. I'll start with dogs. Forget the breeds, a dog is a dog. No, that's not right either; I've only made the stereotype more damaging by extending it to the whole species. Better to say that each dog is its own dog. Of course, I'm still using the stereotypical d-word, but let that pass—is the statement true? Yes. J. D., like any other mutt, is unique; just as each Labrador retriever and each Siberian husky is unique. Is the statement useful? Not very. When selecting a dog to be a companion for a three-year-old child, for example, one ought to know that many but not all Siberian huskies actively resent being chased, poked, pulled, pushed, and kicked, while most but not all Labs regard it as welcome attention. Similarly, it is sometimes useful to know that Alsatians and a few other breeds are prone to hip dysplasia; dachshunds often have back trouble; dalmatians, unlike other dogs and most mammals—but like apes and humans—excrete uric acid in their urine; and that beagles, although intellectually challenged, are exceptionally good at hunting rabbits. Breed stereotypes have their value.

Despite the importance of understanding breed characteristics, it is still foolish to assume that all dogs of a particular breed are alike. A single litter of puppies may contain individuals that are timid, extroverted, loving, or unfriendly. Dobermans, famous as vicious attack dogs, often are trained as gentle companions, used to cheer up patients in nursing homes and hospitals. The interaction between the breed or species and the individual is complex. The Dutch naturalist Niko Tinbergen, who won the Nobel Prize for his studies of animal behavior, helped us understand this subtle interaction. Although he eventually learned how to tell apart individual gulls, a remarkable achievement, it was his discovery of special signaling behaviors common to *all* herring gulls that earned him his prize. Tinbergen showed us that the behavior of the individual is often comprehensible only in the context of the group.

There is a caveat, of course. In *Life is a Miracle*, Wendell Berry reminds us that we always need to consider "the distinction between reduction and the thing reduced."

> To be competent a farmer must know the nature of species and breeds of animals. But the better the farmer, the more aware he or she is of the animal's individuality. "Every one is different," you hear the good stockmen say. "No two are alike." The ideal of livestock breeding over the centuries has not been to produce clones. Recognition of "type" is certainly important. But paramount is the ability to recognize the outstanding individual.

Given this caveat, it is still clear that breeds and species are necessary categories. But what about the broader grouping of exotic species? True, we cannot look at a particular introduced species and tell in advance whether it will run wild in its new habitat and whether it will cause damage if it does. The gingko tree, the beautiful species introduced from China, which I planted in front of my house, does not escape from cultivation. The starling, that was much maligned by my friend Claude, was first introduced from Europe into New York City in 1890 and now occupies much of the continent; although people find it noisy, and it is aggressive toward bluebirds and some other native species, it does consume large quantities of insect pests such as cutworms, grasshoppers, and weevils. Despite these kinds of examples, however, there are more than enough cases in which exotic species have been extremely harmful to justify using the negative stereotype.

The fear of characterizing breeds or species is, I suppose, a side effect of the new and long overdue political awareness of the dangers of stereotyping people according to race, religion, or ethnicity. The first of these supposed analogies, between breeds of dogs and races of people, makes no sense. The reason for this is simple. Dogs have been single-mindedly bred for specific, often bizarre traits and behaviors; people have not. Different human groups do have identifiable physical differences, such as skin color, stature, and susceptibility to certain diseases. But traits such as intelligence, character, aesthetic sensitivity, and spirituality have not been similarly sorted out by race—racists to the contrary. Intelligence and strength of character, for example, are equally valuable to people living in the Arctic, in central Africa, or in northern Europe; these qualities have moved with people wherever they went. We expect and find that they have a similar occurrence among all peoples. The other analogy, between stereotyping alien species and stigmatizing human races, is equally far-fetched. While pejorative generalizations about human races are demonstrably untrue, it is a simple matter to show that gypsy moths, kudzu vines, and Argentine ants are destructive precisely because they are alien species in a new environment.

When political correctness calls attention to the plight of oppressed peoples, it serves a good purpose. For example, the outcry against the "racial profiling" that afflicts African Americans and Hispanics in the United States is entirely proper. But when political correctness is applied to our way of characterizing different kinds of animals and plants, it endangers the legitimate uses to which it is put, and deserves ridicule.

So please don't call me a racist or speciesist if you hear me complaining about the alien European, wooly adelgids that are killing New Jersey's hemlocks, or the introduced Caspian Sea zebra mussels that are causing so much ecological and economic damage in the waterways of eastern North America. And if you run into me while I am walking my dog, I will be happy, if asked, to express my opinions about the relative merits of beagles, huskies, Labrador retrievers, and Andalusian bog hounds.

TEACHING FIELD ECOLOGY

The first thing I did when I arrived at Rutgers in the late summer of 1974 was to plan the courses I would teach. My principal fall course was to be based on one that I had helped teach for a few years at Barnard College: The Natural History of the New York Area. At Barnard, I had learned the subject by accompanying far more experienced colleagues — Tony Warburton, an evolutionary biologist, and Patricia Dudley, an ecologist — on their field trips. Now, in New Brunswick, I had a new teaching partner, Jim Applegate, a wildlife biologist, but I didn't anticipate any changes.

Jim listened to my plans for the course with gratifying attention and enthusiasm. He had only a few questions. "What are we going to call the course?"

"'The Natural History of the New York Area,'" I answered, "or maybe 'The Natural History of New Jersey.' That's what it is, isn't it?"

"Sure. But we already have our course in General Ecology, which you run. That's mostly theoretical, indoor classroom learning. Why not call the new course 'Field Ecology' and design it to let students who have had General Ecology apply their knowledge to the real world? In other words, we want to teach them more than descriptive natural history — they should understand the ecological and human processes that make each place what it is."

This meant a pretty complete rethinking of the course, which I hadn't expected to do, but I grudgingly agreed. Thus began what has become the most remarkable experience of my teaching career.

For the first three or four years, we taught together: two different sections a week, each with the two of us and fourteen students crammed into a fifteen-passenger van for field trips that lasted from 1:00 to

6:30 P.M. From the start we decided that there would be almost no class-room teaching, just field trips, regardless of weather. And so we have wit-nessed the majestic silence of a white cedar swamp in the October sun-shine, have walked the springy, low-tide—bared *Spartina* salt marsh in torrential rain, and have given final exams on abandoned landfills dur-ing snowstorms. (The best technique for taking an exam in the snow or rain is to put a clipboard and paper in a large, clear plastic bag and write inside the bag with a medium pencil.)

Jim and I worked well as a team, the born-and-bred outdoorsman and the suburban-grown theoretician. He could show the students ex-actly why the Pine Barrens vegetation and accumulating litter layer in the new paradise-in-the-pines luxury housing development meant that the whole complex was constantly at risk of being incinerated in a forest fire; I could explain Lewis Mumford's insights about places like that and how such developments failed even when they did not burn. As we walked through the Hutcheson Forest, Rutgers's sixty-acre, old-growth wilderness ten miles west of the college, Jim would point out all the ex-otic plants—Norway maple, Japanese honeysuckle, and *Ailanthus*—that were displacing the native vegetation, and I would discuss the various ex-planations of why exotic species are able to invade and alter the charac-ter of a supposedly protected habitat.

Almost of its own momentum, a theme developed that included and greatly transcended my first idea of a regional natural history course. In Field Ecology we expect our students to become able to enter a strange landscape and, by seeing its plant and animal life, its landforms, and its human artifacts, to describe its history, its present condition and use, and its fate in the foreseeable future.

We start small. On the first day of the course a number of years ago, we wandered around the campus, introducing our methodology to the students, all of whom already had taken General Ecology. At one point, I stopped at a four-foot, bushy Colorado spruce that Jim had previously discovered, and asked them to tell me what it told them. They crowded around the little tree, examining and poking.

"It has multiple trunks," said one observant forestry major, "so it must have resprouted after being burned, or cut, or grazed."

"Not burned or grazed," said another student. "Not here in the mid-dle of campus."

"Right," I agreed. "So tell me in what week of what month of what year it was cut, and at what time of day."

They stared at me blankly. "You've got to be kidding," an exasperated senior muttered. A minute or two passed, then a few smiles appeared simultaneously.

"The week, month, and time of day are easy—it was cut the second or third week of December, after dark."

"That's right. And the year?" They gingerly pulled the sharp-needled stems apart, counting the annual growth intervals back to the point where the trunks diverged.

"1988," they announced after a brief conference.

During this time one student had stayed aloof, a funny expression on her face. I asked her what was the matter. She hesitated, then answered: "I know the fellows who cut it—they were in my freshman dorm. It was a couple of weeks before Christmas, in 1988. At night, so nobody would see. I helped them decorate it." She looked at her classmates in amazement. "I can't believe that you guys figured all that out just by looking at the tree."

From single trees we progressed to whole ecosystems. The presence or absence of wild grape and bittersweet vines, flat versus hummocky topography, invasion or lack of invasion by exotic species such as winged euonymus and *Paulownia*, a strong or indistinct deer browse line, regeneration or no regeneration of tree seedlings, pin oaks versus chestnut oaks, maple-leaved viburnum and white snakeroot versus highbush blueberry and sphagnum moss, caddis fly and mayfly larvae versus *Tubifex* worms and *Physa* snails . . . these were the elements of which stories were made, histories were deciphered.

After a few years, Jim and I stopped teaching Field Ecology together. Two professors for fifteen students twice a week was too great a luxury for a modern research university. I was suddenly on my own, and very nervous. Fortunately, the students were well trained in the subjects I had never had in college: dendrology (the study of trees), geology, soils, wildlife ecology. "Why is this black locust here?" I would ask, pointing to a handsome tree with a strongly patterned bark that identified it to me.

"Er, I think that's a sassafras, sir," said one of the star dendrology students, politely. I didn't argue; after all, she was helping to teach the dendrology course. Besides, I rationalized to myself, it's good for them to see that the professor is fallible. And there was my wonderful memory of Professor Nash, one of my chemistry teachers at Harvard, whose lecture demonstrations were often carefully planned to turn the wrong color, or even blow up, because he felt that nobody learned anything from demonstrations that worked properly. Of course the analogy was backward: I was the one who was learning from the mistakes.

Meanwhile, in the other section, Jim dropped the urban/suburban field trip because he had been uncomfortable with it and added an exercise that I haven't yet had the guts to try. It's called "Stump the Professor." One year, for example, two of his best students brought Jim and the rest of the class to a completely unfamiliar forested area that they had secretly researched in advance—the way Jim and I scout out our trips ahead of time. It took Jim about half an hour to figure out that it had been a dairy farm until the 1940s. Nobody else got it, although the better students came close.

In recent times I have noticed a difference in the training of the students who come on my field trips. They are just as bright as before, but many of them seem to have had little direct experience of the natural world while in college. Here's an illustration. We have a wetland field trip, designed by Joan, that begins with a visit to the floodplain of the Raritan River. I show them that in the floodplain the river bank is highest right next to the river, because during floods the waters slow down as they pour over the bank, dropping their coarsest and heaviest suspended gravels and sands first, thus building a natural levee along the river. As the floodwaters move farther from the river, they slow down even more and deposit a thinner layer of lighter particles, fine silt and clay. The two zones of the floodplain have very different vegetation; the higher, better-drained zone next to the river has many species of trees and shrubs, while the lower, wetter zone farther from the river, which is more frequently flooded (when waters pour through gaps in the levee), has fewer species of trees and almost no shrubs. Recently, at Joan's suggestion, I started bringing along a soil auger so that we could get actual soil samples rather than take the theory for granted.

"Most of you have had Soils, right? The students nodded. "Then you've used soil augers." They shook their heads. That wasn't part of the course. Not their fault, I realized, but it got me upset anyway. Going out to look at the natural world and to learn from it is deemed too expensive by cash-hungry universities (although a miniscule fraction of the cost of equipping a gene-splicing lab), and why spend resources on an unglamorous subject like soils? Needless to say, the student products of this new, tough-minded pragmatism can be woefully helpless and infuriatingly uncreative when let loose in a real environment.

But the field trips are still great fun. Also, New Jersey is probably the best place in the United States in which to teach field ecology. Nowhere else is such a bewildering diversity of mineral types, geological formations, plants and animals and such a continuum of rural-to-urban human settlements crowded into one small area. Within a fifty-mile radius of Rutgers, we can examine a trout stream strewn with boulders in a hemlock-lined gorge. We can visit a holly forest dotted with cactus on a narrow barrier beach. We can look at the amazing differences in vegetation between neighborhoods that were settled by immigrants in the 1930s and those built by developers in the 1990s. Or, in this state whose population density is greater than Japan's, we can explore sparsely settled semiwilderness areas that contained more towns and people 150 years ago than they do today.

Finally, for me at least if not for the students, there is the exhilarating, terrifying, enlightening sense of the power of the passage of time that comes with paying close attention to the same places year after year. Imagine what it is like to show the class lawns, flood retention basins, and townhouses where prior classes have watched potatoes being harvested, and to explain why the only crops they will see on their agricultural field trip are corn, soybeans, and Christmas trees, while their predecessors could also visit orchards and multicrop truck farms. Then erase that image and picture yourself standing with the class on the eroding secondary dune of an ocean beach, holding the faded map of the self-guiding nature trail that you walked with other students twenty years ago. Now imagine yourself telling them that stops 1 through 25 on the trail—the ghost red cedar forest, the gnarled black cherry, the giant poison ivy vine, each with its numbered signpost corresponding to the numbers on the map—are, as you speak, one hundred yards out at sea, like the drowned villages of the Lenape, deep under the waves. Do you feel small? Do you feel transient? Do you feel old?

Here are the two great kinds of change we encounter in life, cultural and ecological, bobbing and weaving together, forever repelling and embracing, inexorably meeting at the myriad points of our own mortality. Field Ecology has let me see this ghostly process in a sharper light than is given to that sad majority that lives—for a while—only in the dim and static world of its own unnatural contrivance. But though it saddens and frightens me to have relentless time made so vivid and so personal, I can yet enjoy watching the stately and immensely complex dance of culture and nature, and can understand a little better the society and place in which I live.

MORE FIELD ECOLOGY:
RIGHTOFWAY ISLAND

On all but one of the field trips in my Field Ecology course, I take my students to the sorts of places that they have been to before: the beach, the pinelands, the Highlands forest, farms, old fields, streams, salt marshes, suburbs. But on the third trip, after the ones to the campus and to the experimental plots at Hutcheson Forest, we go to a place that is, for all its superficial familiarity, altogether different and exotic. This trip is to America's deserted empire, what the person who knew it best, ecologist Frank Egler, described as the Rightofway Domain.

Right-of-way land comprises at least fifty to seventy-five million acres in the United States, an area larger than New England. It is disposed as long strips of property along railroad tracks, roads, and canals; under power lines; and above buried pipelines. Many of these rights-of-way, even those in heavily populated areas, are scarcely ever visited by people—they are cut off from human presence by fences, no-trespassing signs, patrolling police, dense vegetation, and a scarcity of reasons to set foot in them.

True, some abandoned rights-of-way have been put to use. The towpath and adjacent land along the old Delaware–Raritan Canal, which winds its way for many miles through central New Jersey, has become a very long, very narrow, very popular state park. Indeed, this is the park that has helped protect the forest along the Millstone River, which I described earlier. Hunters love the rights-of-way under power lines, which attract deer and small game. And disused railroad lines have been turned into foot and bike trails in several parts of the country. But many rights-of-way, totaling a huge amount of land, go for months or years without feeling a human step or hearing a human voice. These places are themselves neither urban nor suburban nor rural; neither settled nor wilderness. They are a quintessential part of what author James Howard Kunstler has called "the geography of nowhere."

In my right-of-way trip we start with a boggy strip of land above a transcontinental gas pipeline. The students examine the vegetation and deduce that it is maintained by one annual mowing, in the fall. From the pipeline, we go to a railroad line to observe the effects of the powerful herbicides that are sprayed on the plants growing along the tracks by moving tank cars. We used to spend some time walking beside the main-line tracks of the Northeast Corridor, which runs from Washington to Boston. Then one day we were stopped by a large man with a dog, a gun, a rattlesnake belt buckle, and an Amtrak badge. The dog was a well-developed German shepherd, not wagging its tail. The man seemed interested in what our class was doing; he asked questions. Then he told us to go away—and we went. Since then, we have taken a quick, furtive look at the tracks from a protected vantage point behind a factory.

The next stop in our trip is a fascinating brushy strip under a high-voltage power line. The towers that support the power cables dwarf the red maples, sweet gums, gray birches, and pin oaks that have grown up in this damp right-of-way. It takes the students about ten minutes to figure out that the power company fells the weedy trees under the line approximately every ten to twenty years, using both bulldozers and chain saws. I don't like to spend too much time there because of the electromagnetic fields that the cables are generating. Maybe the Swedes and Russians are wrong and the fields are harmless, but why take a chance? Some of the students wonder whether the people living in the apartments next to the power line suffer any ill effects. I tell them I doubt it— the fields quickly diminish with distance, as I show them with my milli-gauss meter. But their children, who play under the power line, may not be so secure.

Naturally, I save the most exciting stop for last. Driving the van for a short distance down Interstate 287, westbound, I tell the students what I know about the planting and care of the right-of-way land that runs along the highway margins. In the first year we did this field trip, Jim and I experimented with pulling over and letting the students get out and look. But the noise of the cars and eighteen-wheelers made it hard for the students to hear us, and the state police were not happy with our being there, so we had to find a better way of teaching about the interstate's right-of-way.

Now, after a mile or two on the interstate, when I get to State Route 27, I take the southbound exit around the cloverleaf, drive on Route 27 under the superhighway, and pull into the adjacent chemical company's parking lot immediately on the other side of I-287. With the students following, I walk across the parking lot to the exit ramp from eastbound I-287. Waiting for a break in the exiting traffic, we cross the roadway and then cross the parallel entrance ramp leading from Route 27 on to the interstate. If you have followed me so far, you know that we are now in one of the four leaves of the cloverleaf. This is Rightofway Island, our destination.

About five acres in area, the nearly circular island has concentric bands of vegetation. Next to the entrance ramp we just crossed, which forms about three-fourths of the island's perimeter, is a narrow zone of grasses mixed with the pink-flowered, ground-hugging relative of clover called crown vetch. The vetch is probably of the variety known as "Interstate," which was developed as a low-maintenance planting along interstate highways. This zone is mowed occasionally. Next to it is another zone, maybe one hundred feet wide, with tall weeds and low-growing bushes. In the middle is a zone of trees. On the far side from where we are standing, the island is bordered for about one-fourth of its circumference by the high, vetch-planted bank that supports the interstate, out of sight but not out of hearing, above.

We move hastily through the weedy zone, stopping only to eat a few of the everlasting peas that have ripened in the September sun. Our goal is the dumbbell-shaped patch of woods at the center of Rightofway Island; I will not be able to relax until we are safely in its shade, not visible from passing cars. The last barrier between us and the woods is a belt of multiflora rose, a thick tangle of intertwined canes studded with stout thorns. But the path we made through it last year and the year before that is still there, and we arrive at the little grove with only a few minor scratches.

The trees of the grove are nearly all sassafras and nearly all are the same age, about thirty-five to forty-five years. They must have sprung from the seeds in berries dropped by birds soon after the interstate, with its cloverleaf, was built. The trees have grown thickly, and the shade has kept the ground pretty free of vegetation, except for an occasional creeper of Japanese honeysuckle where gaps in the canopy let in some light. Evidently this spot never received any fill during the construction of the interstate, so it is a bit lower than the rest of the island. We cannot see out beyond the encircling bushes.

I find a place to sit up against a tree trunk. The students follow suit. Soon, despite the ever-present roar of the interstate and the lesser growl of Route 27, a novel awareness comes to most of them—I can see it in their faces. We are *alone*, they realize, cut off from the civilization around us.

Like Dorothy and Toto in Oz, nobody in the world we have left can see us or hear us—nobody else from our world knows that this place exists. There are no empty cigarette packs or beer bottles on the ground, no footprints except ours; not one living soul has been here since my last visit, a year ago. We are conscious of the world we have left—the noise and the polluted air accomplish that—but, as if we had passed out of familiar time and space, our world is no longer conscious of us.

Here is the New Wilderness, isolated from surging humanity more effectively and completely than the inner reaches of the jungles of the River Amazon. No ecotourists will ever come here. The little grove exerts a kind of weird fascination; it casts a spell, but not the spell of a magical woodland or enchanted garden in a fairy tale. There is a quality of magic in the isolation of this place, but it has an incomplete, purposeless, unsettling feeling. This is Oz with a difference. There is no yellow brick road in sight.

Not much lives in this grove, aside from the trees. There are no lichens on the bark, no snakes on the ground, and few birds among the branches. We have not found any nests. One of the students sees a rabbit—there are probably field mice and voles out in the weeds and vetch. A raccoon or skunk able to make it alive past the encircling lines of traffic would have to run the gauntlet again every time it wanted water. Once, when we were seated among the sassafras trees, a young deer wandered by a few feet away, moaning loudly. But Rightofway Island was not always like this.

In the center of the grove are two mounds of rotting wood. Twenty years ago, they were two dead trees, standing about fifteen feet apart. Then the students could still identify them as apple trees and could understand their significance. There is meaning in two apple trees growing not far from Route 27, since colonial times a part of the main road from Philadelphia to Princeton to New Brunswick to New York. This island was a place where people used to live, in the days before the interstate, before the chemical company, before the strip malls, and before the members-only discount stores. Perhaps the trees were part of an orchard—more likely, they were in the yard of a house. Now, nobody will ever live here again, not even after the interstate highway, transient relic of the age of oil, has carried its last truck howling into the night. The adjacent massive bank of fill that supports the interstate roadway has made this an unsuitable site on which to build, at least until the next glacier scrapes it away. Nor can it, while the highway lasts, return to what it once was, an eastern deciduous forest with a great diversity of plants and animals. Because of the traffic pattern, one could not even put a fast food restaurant or gas station here.

I explain the mounds of decayed wood to my students and ask them to think of the tens of thousands of other pieces of land cut off from people and nature by superhighways. The ecologists Frank Egler and William Niering showed how, at little cost, we can help the land under power lines support huckleberry bushes and quail, but there is nothing comparable that can be done about this kind of place. It is too small and too isolated by unceasing traffic; it has no water. A mere fragment, the island has no connection with any adjacent, larger woodland that could serve as an enriching and rejuvenating reservoir of plants and animals. No squirrel will ever bring an acorn here. It is in limbo, like so many of our works of progress, accessible to neither the creativity of humanity nor the creativity of the natural world. Is this a preview of remnant nature in our midst? A resident alien, cut off from its larger, life-giving self? If so, what is to become of *us*?

The sun is low in the west and it is time for us to leave. Once again we successfully cross the two lanes of opposing traffic. A truck driver honks at the female students; we have left Oz. I look back at Rightofway Island and think that of all the places we visit in Field Ecology, it is the strangest

and most sad. Everyone is anxious to get back to the van. As we drive up
to the gate of the lot, the security guard smiles and waves us on.

A WALK IN THE WOODS

The forest was small by American standards, perhaps fifty or sixty acres,
but in the rolling Devon countryside with its parceled fields and narrow,
hedge-enclosed lanes, it felt appropriately spacious. I was enjoying the
guided tour in one of my favorite parts of the world. The light rain and
chilly July breezes felt right; the English ivy carpeting the ground seemed
right; the leafy hardwoods looked right (although I didn't know the
species and could easily have been fooled); and the probably medieval
bank and ditch running through the woods at right angles to the path
gave everything an impressive air of authenticity. The path turned. As
we rounded the corner, I saw ahead the darker shade of conifers. Soon
we were in the midst of a grove of youthful but already towering Cali-
fornia-coast redwoods. A deep silence hung like a benediction over the
dark wood, but it was quickly shattered. "I'd give anything to be allowed
to cut them down," said our guide, Stephan, in an angry voice.

   This incident passed out of my thoughts until weeks later back in New
Jersey, when I was reading the chapter on conservation in Oliver Rackham's *The History of the Countryside*, an account of the origins of Britain's
landscapes, flora, and fauna. The British landscape of the late twentieth
century, Rackham wrote, is suffering from an acute loss of meaning—
the unique messages once conveyed by many historic woodlands,
witness to millennia of slow and painstaking change, have been garbled
beyond recognition in five or six decades of modern planting, "restoration," development, and agriculture. The more I read in this remarkable
book, surely one of the most profound and eloquent descriptions of
people and nature ever written, the more I understood Stephan's
feeling that his Devon woodland had been desecrated by the planting of
those redwoods. I also began to understand how little I knew about the
long discourse between people and trees in Britain, where the history of

the relationship is probably as well documented as in any place on earth.

As a resident of eastern North America, much of which is wooded, I have known since early childhood what a forest is. It is a place where many trees grow together, a place defined by its trees. In a forest, the human presence is optional and variable. Humans can have a great effect on forests—or hardly any—but we are external to the basic idea of forest. We use the forest, manage the forest, abuse the forest, destroy the forest, protect the forest, enjoy the forest, fear the forest, study the forest, and ignore the forest; always the forest is measured against an original Forest in which people do not appear at all. (This idea of Forest ignores the major role played by American Indians in shaping the structure of North American forests for thousands of years before the Europeans arrived.)

As it turns out, the word *forest* throughout its long history has had a much weaker relationship to trees and a much stronger relationship to people than its current meaning implies. The original Latin word *foris*, from which *forest* evolved, simply meant "outside." In England, the association of forest with trees was even weaker. The word was first introduced in the eleventh century by William the Conqueror; it meant an officially designated place where deer were protected and their usage governed by special laws and regulations. A typical forest might have included villages, towns, agricultural fields, and nonwooded countryside, in addition to patches of woodland. As Rackham points out, the designated forests in Britain were mostly in the less wooded regions, and woods were usually only a small part of them. Some of the largest forests had little woodland; for example, Dartmoor Forest was nearly all moorland and Sherwood Forest was primarily heath. Robin Hood and his merry men must have spent very little time hiding behind trees.

Most official forests were under the jurisdiction of the king; some were controlled by nobles or senior clergy. The forestal rights included primarily the right to take deer (or honorary deer, such as pigs), the right to appoint the sizeable bureaucracy that administered the forest—wardens, seneschals, justiciars, verderers, etc.—and the right to collect fines for violations of forest regulations. A gift of game in honor of a wedding, graduation, pregnancy, or other festive occasion, or the gift of an appointment as under-forester had great political value for a cash-poor medieval king.

The benefits the forests provided, however, included far more than the collection of game and the privilege of appointing administrators. There was also the ownership of the land, the mineral rights, rights of pasture, and, in the wooded areas, the rights to timber and to the underwood, the multiple stems that resprouted from the stumps or roots of cut trees. Rarely did all these rights reside in the same hands. For example, a nobleman might own the land, the king might control the game, timber, and administration, and the commoners often had the valuable rights to the underwood and pasture. While kings frequently quarreled with the nobles over forest privileges, they tended to respect the rights of commoners, which predated the designation of the forests. Not until the enclosures of the eighteenth and nineteenth centuries did many of these forests pass exclusively into the hands of the landowners, who often promptly cleared the wooded portions for cropland.

But many patches of woods, some large, remain in the surviving official forests and in the originally wooded parts of the country that never were in the forest system. This British woodland is not an unchanged remnant of the wildwood that covered the land 6,000 years ago, when the first Neolithic settlers arrived and began, with amazing efficiency, to convert the landscape to open, agricultural fields. Although the kinds of trees once found in the wildwood still exist, the woodland itself is now something different, as much a cultural as a natural environment.

One difference is a striking change in the relative abundances of the various tree species. In the wildwood that spread over much of lowland England, for example, the dominant tree was the stately lime or pry, *Tilia cordata*. (A close relative in the northeastern and northcentral United States is *Tilia americana*, the American basswood, or linden.) We know that the lime was common from the record of preserved pollen grains. Now it is rare over much of its former range, while oak, ash, and beech, which like starlings and raccoons are comfortable around people, have increased at the lime's expense. One reason may be that the lime is particularly attractive to the herbivores that people introduce and favor—deer, rabbits, and sheep; its American cousin, the linden, is similarly sensitive to browsing. There are no doubt other reasons why the lime gets along uneasily with British civilization, but they are hard to fathom. In south Suffolk, Rackham writes, lime is still fairly abundant in a few woodlands that are known or believed to be ancient, but in places where these woods are expanding into adjacent fields, the lime is reluctant to go beyond the old woodland boundaries that were set in Neolithic times. Some flowering plants, such as the oxlip, *Primula elatior,* have the same reticent behavior.

Another critical difference between the wildwood and the woodlands of today is the practice of coppicing—cutting the underwood every few years. Coppicing has been a hallmark of British woodsmanship since the Stone Age and is one of the best examples of sustainable forestry anywhere. Underwood has had many structural uses and has been a major source of fuel. When done properly, coppicing can continue indefinitely; Rackham notes that the Bradfield Woods have been cut at least seventy times and are thriving. An ash that would normally have a lifespan of 200 years can live and be healthy for more than a millennium if coppiced, its stump, or "stool," spreading in rings that are as much as 18 feet across. In wooded areas that were heavily foraged by game or cattle, the valuable resprouting stems were likely to be eaten, so trees were frequently pollarded—cut at a height of 8 to 12 feet above the ground. This put the new growth of stems above the reach of browsing animals. Pollards were repeatedly cut for wood, like coppice.

A woodland with a mixture of coppice and timber is not only diverse and attractive, but is also immensely durable, yielding products of great value century after century at minimal cost to the people who use it, while retaining the species of plants and animals that give it its distinctive and stable character. This is why much of the British woodland extant in 1700 would have been familiar to an Anglo-Saxon resident in the year 1000.

Since 1700, many acres of woodland have been lost, some to plantation forestry and ill-conceived restoration projects, but mostly to agriculture—first through the enclosures and recently through the rise of industrial agribusiness. The conversion of woodland to agricultural fields has been, overall, both a social and an economic failure. Because woodland usually is found in places unsuitable for agriculture, and because modern industrial farming is vulnerable to disturbance and expensive to maintain, the replacement of woods by farms has yielded little or no lasting benefit to their owners.

In the remaining woodland, the art of coppicing has also fallen on hard times. At present, oil is easier (but not cheaper) than underwood to use as fuel; other materials have replaced coppiced stems; timber is readily available from Canada; and forest foods such as hazelnuts are hardly a necessary part of the British diet. Yet coppicing has not died; in the 1990s it was revived in a conscious effort to maintain historic practices. It is a source of firewood to be used instead of oil, provides materials for special crafts, and creates animal and plant habitat. Whether coppiced woodlands, one of the oldest and most successful compromises between people and nature, will have a new role to play in the coming era of scarcities remains to be seen. It is not out of the question.

In the meantime, woodlands still grow all over Britain, and some men and women still know how to care for them. I take it as an encouraging sign that there are people who can rage against the planting of a redwood grove in a Devon wood, who can resent and resist the breach of faith and loss of meaning that mark so many of our dealings with the landscape. Long in the making and long in the proving, the ancient compact between Britons and the natural world may yet survive the uncertain centuries ahead.

DEGREES OF INTIMACY

When we arrived in Vancouver at the start of our vacation, the tabloid headline at the newspaper stand caught our attention. "World's Bravest Mom," it shrieked. We stopped to read. The story was simple; it needed no journalistic embellishment. Dusk, August 19, 1996. Mrs. Cindy Parolin is horseback riding with her four children in Tulameen, in southern British Columbia's Okanagan region. Without warning, a cougar springs out of the vegetation, hurtling at the neck of one of the horses. In the confusion, Steven Parolin, age six, falls off his horse and is seized by the cougar. Mrs. Parolin, armed only with a riding crop, jumps off her horse and challenges the cougar, which drops the bleeding child and springs at her. Ordering her other children to take their wounded brother and go for help, Mrs. Parolin confronts the cougar alone. By the time rescuers reach her an hour later, she is dying.

The cat, shot soon afterward, was a small one, little more than sixty pounds. Adult male cougars can weigh as much as 200 pounds, we learn the next day from the BC Environment's pamphlet entitled "Safety Guide to Cougars." We are on our way to Garibaldi Provincial Park, where we plan to do some hiking, and have stopped in the park head-quarters for information. "Most British Columbians live all their lives without a glimpse of a cougar, much less a confrontation with one," says the pamphlet, noting that five people have been killed by cougars in British Columbia in the past hundred years. (Actually, the number is now higher; cougar attacks have become increasingly common in the western United States and Canada in recent years.) "Seeing a cougar should be an exciting and rewarding experience, with both you and the cougar coming away unharmed."

However, the pamphlet notes, cougars seem to be attracted to chil-dren as prey, possibly because of "their high-pitched voices, small size, and erratic movements." When hiking, "make enough noise to prevent surprising a cougar . . . carry a sturdy walking stick to be used as a weapon if necessary," and "keep children close-at-hand and under con-trol." If a cougar behaves aggressively, face it, speak loudly and firmly, wave your stick, act threateningly, fight back if attacked. Make eye con-tact with a cougar, the ranger behind the desk advised; on the other hand, never look straight at a grizzly bear—it may be provoked to charge. This year, he added as we turned to go, we've had a very cold, wet spring, and that has brought the cougars and bears down into the more frequented parts of the park. We should take reasonable precau-tions but enjoy our hiking. As he spoke, I thought of Mrs. Parolin, who had done all the things that the pamphlet recommends.

Arriving in the town of Whistler, I visited a few shops in search of a suitable walking stick, but most of the sticks I saw looked like high-tech ski poles and wouldn't have fought off an angry cricket—the one wooden stick I found was flimsy and cost $55. Besides, Joan was plainly amused at my quest, and that took the pleasure out of Yuppie-ish shop-ping. God or nature would provide a walking stick, I decided, with a final, backward glance at the shop windows.

The next morning we got to the trailhead early. The sun was bright above the horizon—and there was a bark free, sturdy tree branch, just the right length, that someone had left on a flat stone at the beginning of the path. We set out. Sam and Jon had been told not to run on ahead as they usually did. And for the first time in our hiking history, we asked them to make noise, even to argue if they felt like it.

The trail wound steadily upward. As soon as we crossed the park boundary, we passed from second growth to mature forest, whose majestic conifers were draped with lichen strands that caught the golden sunshine in ten thousand ripples of light. The boys were totally silent and I found myself thumping the ground vigorously with my stick to compensate. But the tiny noises I made were swallowed up in the immensity of the forest, which had its own sounds. There were Flute and Oboe Creeks, for example, which we crossed on small wooden bridges, rushing headlong down their valleys in a froth of wild, musical energy. And there were gray jays, looking, as the Peterson field guide says, like huge, overgrown chickadees, chattering loudly away as they hopped onto the boys' hands to receive cookie fragments. We met no cougars or bears, but once a vole scuttled across the path ahead.

Gradually, as we ascended, patches of bright, subalpine flowers appeared in openings in the forest. Then the patches coalesced into a dazzling, waving carpet of yellow, white, and purple, and the land opened out into a saddle between two peaks. We were in Singing Pass. The trail continued on ahead, switchbacking up the far peak, and we followed it, climbing above the tree line, past and across fields of late summer snow, which was excellent for snowballs, round a talus slope of rocks and boulders, to a point where we could look down on Russet Lake, below, or raise our eyes to a panorama of snow-capped peaks and the brilliant gleam of an arm of Overlord Glacier away in the east. We had come six miles. It was time for lunch.

The return trip was also uneventful. I pounded on rocks and roots with my stick, enjoying the totally different views from the trail now that we were going downhill. Again, there was not a sign of cougar or bear, no distinctive footprint with three lobes at the base of the pad, no rotting tree torn apart in a search for grubs. At the parking lot, I put the stick back on the flat rock. It had been totally effective in warding off danger, and I was happy to leave it for the next group that wanted to walk to Singing Pass.

The rest of our trip also passed without incident, although not without wonder. We saw old-growth, western red cedars more than twelve feet in diameter in the rainforest on Meares Island. They had been saved from the chainsaws of MacMillan Bloedel by the joint action of the First Nations of Western Vancouver Island and a coalition of North American environmentalists. Their peaceful protests in 1993 resulted in 865 arrests, the biggest act of civil disobedience in Canadian history. We saw magnificent coastal bogs with dwarfed yellow cedar and lodgepole pines, and with sundews like those in the New Jersey Pine Barrens. We watched a bald eagle feed a fully fledged offspring that was bigger than she was. But the largest carnivorous mammals we came across were three raccoons foraging at a dumpster in downtown Vancouver, two blocks from Stanley Park. No cougars at all. Not surprising, said Neil, one of our Canadian biologist friends. They had tracked an adult cougar that was wearing a radio collar for eight months while it was living in and around a suburban development, eating cats and dogs. During that entire time, it was not once seen either by the tracking crew or by any of the residents of the community. We didn't see any cougars; how many saw us we will never know.

A week after our return, I had a visit in my office from Mark, one of Joan's and my students who had just gotten back from Professor Roger Locandro's celebrated ecology class in Alaska. Mark and another student named Anna had gone for a few days of camping in Denali Park before the class started. He told me a hair-raising story about their trip. While walking alone on a ridge trail near their campsite, he heard rustling in the brush behind him; out onto the trail came a large Alaskan brown bear, followed shortly by her two, sizeable cubs. They were about twenty feet away. Nearly petrified, Mark began walking slowly backward down the trail. The cubs pushed on ahead to investigate him, but were held in check by a peremptory "Woof!" from their mother. Slowly Mark retreated and slowly the bears advanced. Mark realized that he had to get off the trail; the only practical alternative was a shale slope which ended in a precipitous drop-off about forty feet from the path. Stepping down onto the shale, he promptly fell and slid halfway to the edge, only stopping himself by wedging his foot against a rock. The cubs also left the trail and started coming down the slope toward him. They were halfway down when their mother's "Woof" stopped them in their tracks—another woof brought them quickly back up to her.

As Mark lay stretched out on the shale, he looked up and saw the 600-

pound mother grizzly looking down at him from the path. She was an amazing golden color, and he was intensely aware of the minute patterns and matting of her fur. He remembers that the thought of how incredibly beautiful she was filled his mind, and her great calmness overwhelmed him.

He waited ten minutes after the bear and her cubs moved on, then scrambled up to the trail and found his way back to camp. When he arrived, Anna saw that he was deathly white and his eyes were as big as saucers. She told him that while he was away she had been singing a little song—perhaps a prayer: "Bears, please don't eat us. Bears, please don't eat us." Since that time, Mark said to me, many things in his life seem less important than they used to. He feels deeply changed.

There are degrees of intimacy with nature, even if we are not always aware of them. For most of us who love wild nature, our contact with it, limited by our particular civilization, is at the surface, and our little forays only give us glimpses and hints of the immense, rolling power that lies beneath. But occasionally one of us suddenly, without warning, slips below the surface and there meets vast, impersonal forces and the probability of terrifying change—or death. Only a few seek this degree of intimacy deliberately; many of them do not survive. And yet every now and then there is that charmed individual who, against all common reason and experience, passes in and out of deepest intimacy with nature at will. I think of John Muir climbing a wildly swaying, 100-foot Douglas fir in a High Sierra windstorm, only coming down hours later after the storm was over, and discovering in the ensuing calm that "never before did these noble woods appear so fresh, so joyous, so immortal."

FIVE

RESTORING THE COMMUNITY

Good forms confer health upon the things that they gather together. Farms, families, and communities are forms of art just as are poems, paintings, and symphonies. None of these things would exist if we did not make them. We can make them either well or poorly; this choice is another thing that we make.

Wendell Berry, *Life Is a Miracle*

THE UTOPIA FALLACY

Pray for the welfare of the [ruling] kingdom, for were it not
for the fear of it, men would swallow one another alive.

Pirkei Avot (Ethics of the Fathers) 3:2

Also a common Arabic proverb

Utopia: An ideally perfect place, especially in its social, political, and moral aspects. . . . Greek *ou*, not, no + Greek *topos*,
place.

The American Heritage Dictionary, 3d ed.

Confrontation with a grand and formidable opponent can produce par-
adoxical results. It can unite in a common opposition people who have
nothing else in common. It can evoke dramatic reserves of courage, energy, and creativity where none were suspected to exist. It also can cause
sensible people who should know better to act as if the resolution of
today's problems will inevitably usher in a golden age tomorrow, as if the
ending of one evil could not possibly lead to another. This is the Utopia
fallacy.

The central problem of our age has been bigness: bigness of corporations, weapons systems, trade organizations, human population, media
conglomerates, and—enabling all other bigness—the power of science
and technology. Only the earth itself is large enough, barely, to serve as

a comparator for the scope of our undertakings—hence the ugly term *globalization*, used to describe the processes that have increasingly ruled our lives during the past century. So relentless has been the advance of bigness, so rapidly has it swallowed up everything that is small and self-contained, that it seems about to engulf the entire world, which will be fixed forever in the grip of one global production system, one global trade organization, one global currency, one global store, one global language and culture, one global media outlet, one global selection of foods, one global political entity, and one global police force. Once this condition has been reached, historical change will cease and human freedom will be a thing of the dead past. Such is the fear of many. But terrifying as it may be and useful as it is as a rallying point for the fight against globalization, this particular vision of the future is more threat than substance.

Bigness and globalization are nearing the end of their dominion. For a while there will be more conglomeration and centralization, and a decade or two or three of additional damage—but the days of globalization are numbered. Now is the time to prepare ourselves for the world that will follow, a world of smallness and decentralization.

It may seem premature to anticipate the demise of globalization while giant corporations are merging or swallowing each other up at a frightening rate, and new technologies that shrink the earth and enhance our ability to change it are announced almost daily. Yet remember the Soviet Empire in 1970: the tightness of its control over each and every individual and nation in its vast domain was unprecedented in human history. Who would have predicted that it would all be over twenty years later? The philosopher of history John Ralston Saul has written, "Nothing seems more permanent than a long-established government about to lose power, nothing more invincible than a grand army on the morning of its annihilation."

The communist Soviet Union paid the price for expanding its power and control without heeding the limits of tolerance of its human and natural resources or the dangerous failures of its technology. Today, with the world dominated by the capitalist techno-corporate system, the same theme is being played out again. Information is distorted and knowledge lost as they are manipulated in the service of greed and power; the poorly understood complexity inherent in centralized control enhances the potential for massive mistakes and accidents; resources are being exhausted and the earth's natural regulatory mechanisms disrupted; and belief in durable values and any sort of higher inspiration is being steadily undermined. The effects are everywhere: floods of refugees; the proliferation of local wars; the resurgence of old infectious diseases and the emergence of new ones; serious problems with energy reserves and agriculture; global warming and rising sea levels; and even in America, for most, a decline in the standard of living and the quality of human services, a growing feeling of personal helplessness and lack of confidence in democracy. The prevailing system produces these problems as a condition of its nature; it is unreasonable to expect that it can resolve them as well. This is why general collapse seems inevitable.

Meanwhile, as the social and environmental damages mount, the protest against bigness and globalization gains supporters. It now brings together environmentalists, labor unionists, indigenous peoples, advocates both for human rights and animal rights, community preservationists, Third World representatives, religious leaders, foes of commercialization, and others of like spirit. I count myself as part of this group. We see ourselves as actors in a global morality play—the forces of good arrayed against the forces of evil. Such a simplistic distortion of an enormously complex set of issues is probably a necessary feature of any clash of world views. The other side has all the power; we, however, have right and truth on our side. This is the familiar and comforting moral stance of revolutionaries. But even though, as I believe, it is largely true that unlimited globalization and the technology that promotes it are amoral and wicked, it is far from being completely true, and there's the rub. The danger of an absolute moral stance is that when the conflict for which it was designed is over, the "victors" may find themselves poorly prepared to deal with the reality around them. For when the good eventually triumphs, in large part as the result of factors that we had nothing to do with, what then?

After globalization comes to an end, are we likely to find large numbers of healthy communities well-stocked with public-spirited, self-sufficient women and men, communities in balance with other, similar communities, all contained within ecologically reasonable bioregions? Will we have enough good, wholesome food to eat, snug houses, roofs that don't leak, and the health to enjoy these blessings? Will we find, as

Frodo did at the end of *The Lord of the Rings*, "a sweet fragrance on the air . . . the sound of singing . . . over the water," and see "the grey rain-curtain turned all to silver glass and . . . rolled back," revealing "white shores and beyond them a far green country under a swift sunrise"? Certainly not. Frodo was leaving the world in which he had struggled so long and to such good effect. We will have no such choice.

We have to stay put, and it will not be Utopia. What we will find is that the damage already done to the earth and its people, plus the hidden costs of the ending of the present techno-economic system with both the bad and the good that it contained, will have conspired together to deprive us of the world of our dreams. For once the problem of bigness and globalization has faded or disappeared, other problems will take its place, just as the nineteenth-century scourge of child labor was followed by the twentieth-century scourge of idle teens haunting the street corners, drug-intoxicated and violent. Many of the troubles of the decentralized, low-tech world to come are not hard to predict, and this is the time, during the tumultuous transition, to think about them if we can. In an address delivered to the International Academy of Philosophy, meeting in Liechtenstein in 1993, Aleksandr Solzhenitsyn said: "We all see and sense that something very different is coming, something new and perhaps quite stern. No, tranquillity does not promise to descend upon our planet and will not be granted us so easily." The new problems, I believe, will sort themselves out into two overlapping groups: techno-scientific and environmental problems, and human social and economic problems.

It is too early to assess the extent and permanence of the many kinds of damage done to the earth in our time, nor can we gauge the effects of this damage on our successors. During the first centuries of European colonization in North America, there was rich, uncontaminated topsoil almost everywhere; abundant, easily accessible oil and iron; and apparently limitless old-growth forests containing massive trees beyond number. In the twenty-first century, these resources are sadly depleted. To fill the resource gap, we have turned to other, poorer and economically defenseless regions of the world, stripping them bare; the harm done to these places is greater than what we have inflicted on ourselves. How will we manage with what is left? A society of rich "consumers" is about to become a society of poor scroungers: ripping shingles, plumbing, and fixtures from abandoned suburban houses, searching the innumerable rubbish heaps left behind by a spendthrift culture for yesterday's trash—tomorrow's necessities.

Some of the myriad concerns that the residents of Utopia will face are scarcity of clean drinking water, the storms, droughts, and floods brought on by grievously disrupted climates, toxic dumps everywhere, the widespread disappearance of useful, practical skills and knowledge, the spread of destructive exotic species, and poorly controlled infectious diseases. Ironically, all of this will happen in the absence of the powerful, globally coordinated science and technology that might have been counted on to give at least reassurance and, occasionally, real assistance.

And what will become of the countless inventions and services created during the technological explosion when runaway technology and its too-permissive parent, science, have been subdued? What will survive and what will disappear? Many of us will not miss supporting the army of technicians and technocrats working on an antiballistic missile defense shield that is almost certainly doomed to fail. Nor will we long lament the demise of NASA's hideously expensive interplanetary program or even the space shuttle. These are technologies that eat up a nation's wealth while offering nothing commensurate in return, and their production requires a level of centralization, coordination, and control unique to history. In like spirit, we will gladly wave goodbye to recombinant bovine growth hormone, along with the terrible and unnecessary social, medical, and economic problems it has spawned. But there are technologies we may be more reluctant to lose. Computers, peripherals, and computer networks consume an inordinate amount of electric power to run and fresh, clean water and other resources to produce—far more than may be available if the majority of the world's people decides to use them. How much of this technology will we be willing to do without? Similarly, we ought to wonder and worry about what it might mean to give up magnetic resonance imaging (MRI), a benign and useful but resource-consumptive technology that depends for its existence on a high order of economic centralization, exploitation, and power. Many other technologies in this category seem equally vulnerable.

Perhaps under ideal circumstances we will be able to selectively save some of the softer, more beneficial technologies for use in our decentralized green Utopia. More probably, the entire technological system is so heavily interlinked and so thoroughly dependent on political and economic props and subsidies, on a steady infusion of wealth and resources, that it will disintegrate almost completely when roughly handled, like a castle made of sand. Other cultures before ours have lost their technologies. According to the ecologist Jared Diamond, the Moriori of the South Pacific, a Polynesian people, lost their agricultural technology and reverted to a hunter-gatherer culture and a simpler political organization when their ancestors settled the cold Chatham Islands and found that their tropical crops would not grow. A perfectly satisfactory electrochemical fax communication system set up between Paris and Lyons in 1860 was closed down in 1865 for lack of public interest. Have no illusions about the permanence of our particular technology.

We should have few illusions about society in the postglobal world. Will we be able to build or maintain just communities with the resources remaining? Without the wealth produced by exploitation elsewhere, will we still be able and willing to afford the pursuit of tolerance, democracy, and justice that we see as a necessary condition of our present lives? Furthermore, a lesson we ought to have have learned from nearly a century of Marxist theory and practice in Russia, China, and Cambodia is that political and social systems do not bring about fundamental changes in human nature. That is not to say that some systems are not better than others in certain respects; Marxist Cuba has demonstrated, for those who are looking, the exciting possibilities of science and technology in the service of low-chemical, ecologically prudent agriculture. But after globalization has ended, after corporations have shrunk to proper size and jurisdiction, after the failure of biotechnology in agriculture with its genetically engineered food and animals, after the industrial patents on living organisms have been revoked, and after the high priests of science and medicine have been unfrocked—when the human population has stabilized or become smaller, and semiautonomous communities have again proliferated, we may wonder why so many people are still suffering.

A major cause of suffering will be problems arising from excessive loyalties to small communities and states. We can expect to see all the ancient, familiar side effects of smallness and decentralization: bigotry, racial and religious hatred, the subjugation of women, and intolerance for handicapped people, homosexuals, visionaries, and anyone else who can be labeled as different.

E. F. Schumacher, the brilliant apostle of businesses and communities built to a manageable, human scale, defended "Balkanisation," the proliferation of small nation-states. Look, he wrote in *Small Is Beautiful*, at the example of Germany, which before Bismarck was an assortment of weak principalities and afterward a mighty (and dangerous) power. His implied question was, which is preferable? But in honesty, those of us who applaud the message of Schumacher cannot help but notice the counterexample of today's Balkans, including Bosnia, Serbia, Kosovo, and Macedonia. Always lurking in the background in a parochial world is the threat of local or tribal war. This is not necessarily the whiz-bang war of smart bombs, "stealth" aircraft, and super-weapons too costly to risk losing in combat. Rather, these are the wars that the military historians call "low-intensity conflicts," the little, local wars of rifles, cheap land mines, truck bombs, box cutters, makeshift uniforms, and throats cut in the night. Low-intensity conflict is as terrible as any war, and it is spreading rapidly, fed by regional animosities and the muscle-bound weakness of the great powers. It, too, belongs to the future, a black shadow over the green Utopian countryside.

Every age has its menace. We ought not fall into the revolutionary's error of assuming that our only task is to slay *our* dragon, the dragon of centralization, bigness, and globalization. For the dragon is not entirely wicked, and many helpless people and unspoiled places are under its protection. Moreover, the very presence of this dragon has kept other dragons in hiding; they will emerge after our menace is gone. As Isaiah Berlin wrote about the ideals we pursue, "Some of the Great Goods cannot live together. That is a conceptual truth. We are doomed to choose, and every choice may entail an irreparable loss." Although we have no alternative but to oppose what we know to be wrong in our days, we had best reserve some of our creative energies to help prepare those who come after us to cope with the less desirable consequences of the end of the old system. In an earlier age, this is exactly what Madison, Hamilton, Jefferson, Adams, and the other framers of the U.S. Constitution and Bill of Rights did—with considerable success. They were not Utopians. They built into these admirably concise documents the freedom for subsequent generations to alter them, for better or worse, according to the necessities of the times. Perhaps the original model for their far-sighted flexibility was the Law of Moses set forth in detail in the Hebrew Bible, but with the specific instruction that it was to be interpreted and applied by the judges of future ages in accordance with the needs and conditions of their times. Similarly, America's founders knew that even after the period of colonial domination was over, there would always be other challenges for Americans to face. That they did not predict all of them does not in any way tarnish their achievement. We too must be effective realists; the starting point of realism is to be honest about the changes that the end of globalization will bring to us all.

The great drama of which we are a part is inevitably a tragedy; the facts of our limitations and mortality make it so. Nevertheless, there are some comforts at hand. Although we have to do our best to bring some initial order to the new world we have struggled and worked for, it will not be our burden to fight the battles that will follow. Tolkien wrote in *The Return of the King*, "Other evils there are that may come. . . . Yet it is not our part to master all the tides of the world. . . ." Still more comforting is the simple truth that throughout the long years of history, great systems of wickedness and wrongdoing have sooner or later always failed, and most likely always will. Whether this oft-repeated breakdown of evil is an indication of the existence of some Design beyond our fathoming or is just the bright side of human fallibility does not matter here. The point is that systems that are greatly at odds with justice and goodness do not last forever. If there is any Utopia in the world, this is where it is.

TRADITIONS

The Jewish holiday of Sukkot has set me thinking about ritual traditions, wondering why they assume their curious forms and how they can outlive their originators and persist for such a very long time. Sukkot, or the feast of Tabernacles, more than any other religious holiday I observe, is defined by odd practices that make no obvious sense, yet they have been performed in approximately the same way every autumn for at least twenty-five hundred years. Although many of my friends feel that tradition sanctioned by religious authority is quite sufficient reason for observing the Sukkot rituals, I find it hard to eat a festive meal outdoors in October in a cramped and flimsy three-sided booth incompletely roofed with cornstalks, without asking myself, "Why am I doing this?" The answer, "Because your biblical ancestors spent forty years in the wilderness living in temporary booths or tents," is adequate up to a point, but it doesn't help to explain why, during six of the seven days of the festival, I am obliged to take four species of plants, a citron and an immature palm frond bound together with two branches of willow and three of myrtle, and point them successively east, south, west, and north, then down toward the earth and up toward the sky.

The scriptural origin of the four species used during Sukkot is simple enough. In Leviticus 23:40, instructions are given that beginning on the first day of the holiday, you shall take "the fruit of goodly trees, branches of palm trees, and boughs of thick trees, and willows of the brook, and you shall rejoice before the Lord your God seven days." No wasted words, no explanations. It was left to postbiblical rabbinic interpretation to fill in the needed details and standardize the ritual. Thus "the fruit of goodly trees became the citron (*etrog* in Hebrew), and the "boughs of thick trees" were declared to be myrtle branches. The method of attachment of the willow and myrtle sprigs to the palm frond (*lulav*), using strips of palm, was specified by the rabbis, as was the notion of waving the species in different directions.

The word *sukkot* (singular *succah*), which gives rise to the name of the holiday, means booths or tabernacles in Hebrew; their biblical mention in association with the seven-day festival comes two verses after the description of the four species in Leviticus. Again, the instructions are terse, saying only "you shall dwell in booths seven days"; again, it was up to the rabbis to decide what a booth is. It turns out to have at least three sides and to be roofed with material cut from plants that grow from the ground; in modern practice this means cornstalks, bamboo poles, or prunings from many kinds of trees and shrubs. The roofing material must be spaced closely enough to provide more shade than sun, but not so closely that it keeps out the rain or obscures all the stars. A succah should not be built under a tree or under the overhang of a house or other permanent structure.

Sukkot was the outstanding Jewish festival in antiquity, so much so that it was simply referred to as *the* holiday. In late biblical times, it seems that the form of the tradition may have varied according to circumstances. In chapter eight of the book of Nehemiah, which describes events in Jerusalem in the fifth century B.C.E., following the return of some of the Jews from Babylonian exile, there is an account of the revival of the festival of Sukkot after a lapse of decades. Upon hearing a reading of the Law, presumably Leviticus, the residents went out to the hills to "fetch olive branches, and pine branches, and myrtle branches, and palm branches, and branches of thick trees, to make booths as it is written." We find no mention of either the willow or the citron, while the olive and, if the translation is correct, the pine have been added. Evidently the Jews at that time identified the four species somewhat differently than their descendants did and do not appear to have waved them, but rather used them for the construction of the festival booths.

The question of which species were meant in the original reference in Leviticus persisted into rabbinic times. Approximately 900 years after the festival celebration described by Nehemiah, the rabbis of Palestine were telling a *midrash*, or homiletic story, in which King Solomon, the wisest of men, persuasively argues that the biblical account of the four species could refer to other plants than the citron, willow, myrtle, and even palm. Many trees bear "goodly fruit," Solomon points out—hence why the citron?—and the "boughs of thick trees" could well mean olives, not myrtle. As for the "willows of the brook," well, all trees require water. By attributing this argument to Solomon, who had been dead for a good deal more than a millennium, the rabbis were able to air their own suspicions about the identity of the four species in a politically neutral way. Having expressed their doubts, they effectively put them to rest by falling back on tradition and authority. "Who explained to Israel that the four species referred to are the citron, palm, myrtle, and willow? The sages; as it says [Proverbs], 'they are exceeding wise.'"

Thus a tradition flows through the centuries like a river, now running a straight course, now picking its way along a braided channel with side streams that peter out or perhaps rejoin the main flow at some point downstream. Meanwhile, the explanations of the tradition accumulate, like flotsam in the river, indicating as much about the character and beliefs of the communities that create the explanations as they do about the nature of the tradition itself. For example, in a modern interpretation, the tradition began during the days of Solomon after the construction in

Jerusalem of the First Temple. The festival, newly postponed until the ingathering and processing of the summer fruits and vegetables and the wine-making were finished, was lengthened to seven days, and centralized in the city. The booths, in this explanation, were temporary housing for the many thousands of pilgrims in Jerusalem for the holiday. And in the twelfth century, long before the days of sociological history and cultural anthropology, Maimonides, echoing the earlier beliefs of Philo of Alexandria, said that the booths were to remind the rich, in the midst of their festivities, of the tribulations of the poor.

Like the booths, the four species are at the heart of Sukkot. Even during the desperate second-century war against the Romans, the Jewish commander Bar Kokhba made sure to supply these plants to his soldiers so they could observe the holiday. Naturally, given their importance, many people have tried to uncover the reason why the citron, palm, myrtle, and willow are used. Lacking an obvious biblical meaning, the four species have accumulated more commentary than the tradition of the booths. The fourth- and fifth-century Talmudic rabbis provided a number of widely varying explanations. In one, the species represent, by their shapes, parts of the human body. Thus the palm frond is like the spine, the citron is like the heart, the myrtle leaf resembles the eye, the willow leaf is like the mouth. Another explanation states that the four species stand for the four matriarchs: Sarah, Rebecca, Rachel, and Leah.

Maimonides, whose matchless understanding of tradition was complemented by his brilliant, rational imagination, saw the four species as symbolic of the Jews' entry into the promised land of Israel: "To remember this we take the fruit which is the most pleasant of the fruit of the land, branches which smell the best, most beautiful leaves, and also the best kinds of herbs, that is, the willows of the brook." Ever practical, Maimonides also points out that these species were common and therefore readily available, and that they would all keep fresh for seven days.

Many of the explanations of Sukkot, especially those given by European rabbis who had no personal experience of the climate and ecology of Israel, seem extremely far-fetched today. But early records from the region itself make clear that the holiday and the four species were long known to have an intimate connection with rainfall and water. This is evident from Christian as well as Jewish sources. In John (7:37), it tells how Jesus, on the seventh and last day of Sukkot, called out to those who thirst—an obvious reference to the custom in that era of pouring water every day of the festival.

Arthur Schaffer, an Israeli agricultural geneticist well versed in both ecology and the historical roots of Judaism, has put together a convincing case to show that Sukkot was a water festival even more than it was a harvest celebration. Writing in a magazine appropriately called *Tradition*, Schaffer reminds us that the Israeli climate has only two seasons, a cool, wet "winter" and a hot, rainless "summer." The wet season begins in September or October, shortly after Sukkot, and continues through April. The actual amount of rain that falls varies greatly from year to year. Prayers for rain begin on the Eighth Day of Assembly, Shemini Atzeret, which follows the seventh day of Sukkot, and they are said daily until the spring festival of Passover. It is commonly assumed that these prayers are not recited during Sukkot lest they bring forth a downpour on the celebrants in their unprotected booths. Nevertheless, all of Sukkot can be seen as a preparation for the invocation of rain. In the words of Rabbi Akiba, a second-century sage martyred by the Romans, "The Holy One, blessed be He, said Pour out water before Me on Tabernacles so that your rains this year may be blessed."

Above all, says Schaffer, the four species of the Sukkot ritual represent the life-giving power of water in a desert land. The Jerusalem Talmud, completed about the year 400, quotes the words of Rabbi Eliezer, "These four species grow on water therefore they come as intercessors for water." A similar comment can be found in the Babylonian Talmud, compiled about a century later. Schaffer explains that each of the species typifies a different ecology of water. The willow, a universal symbol of water, is found in streamside habitats, especially in river valleys. The myrtle grows in riverine thickets and along the slopes of stream banks on steep hillsides and mountains. The date palm is a resident of desert oases, where water bubbles up out of the ground under the rainless sky. And the citron, hardly the most delicious or useful of Middle Eastern fruits, represents irrigated landscapes, requiring more water than almost any other fruit grown at the time. Referring to the "fruit of goodly trees," the Babylonian Talmud makes a telling comparison of the Hebrew word for goodly—*hadar*—and the Greek word for water—*hydor*.

What makes traditions persist for millennia? The social historian Eric Hobsbawm has said that the modern era is replete with "invented traditions," nearly all of them related "to that comparatively recent historical innovation, the 'nation,' with its . . . nationalism . . . national symbols, histories and the rest." Thus we have flags, anthems, contests, costumes, and institutional rituals, mostly used to legitimize authority and particular value systems, or to symbolize membership in a group. Can such traditions persist? Hobsbawm does not elaborate, but common sense suggests that when a nation dies or radically changes, its defining traditions will also die or radically change.

The traditions of Sukkot and the four species have outlasted many nations and empires; they persist today among Jews who live in Russia, North Africa, Iraq, Israel, Australia, India, and America, most of whom look and behave like the other residents of their respective countries. Why are these traditions so durable? One reason stands out in my mind. They are *organic* traditions—arising out of the soil, the vegetation, the landscapes, and the climate of the land that gave them birth. An ecologically rooted tradition, it seems, can have the strength to withstand the passage of time, the rise and fall of dynasties and cultures, and even, somehow, transplantation to a different place on earth.

What are we to make of the explanations of our traditions? I come back to the metaphor of flotsam on a river. It can tell us much that is important to know about where the river has been and what it has encountered, but it does nothing to alter the river's flow and course. For that, other, deeper factors are at work. As the rabbis of the *midrash* about Solomon taught, intellect is good and necessary, but in the end a tradition has a force and mystique of its own, which one can only choose to reject or accept.

Traditions help keep communities together. The old, organic traditions, especially those that are environmentally connected and socially benign like the traditions of Sukkot, are worth maintaining for that priceless function alone. But what if traditions have been lost during centuries of turbulence? All traditions must have a beginning; why not invent new traditions if the old ones are gone? If they are created skillfully and carefully, they will develop their necessary attachments to the communities and the environments in which they must live. Here we see the wisdom of African Americans in inventing and accepting the new tradition of the Kwanza holiday, celebrated in that auspicious time for festivals, between December 26 and January 1. Long may Kwanza survive—together with Sukkot and all other community-sustaining traditions—so that some day its origins, too, will be a subject of debate and speculation.

JANE AUSTEN AND THE
WORLD OF THE COMMUNITY

For two weeks now, I have wallowed in sinful luxury, rereading the six completed Jane Austen novels (especially my favorite parts), basking in the warmth and wit of her collected letters, eagerly absorbing the details of her life from her best biographies, and attentively following the arguments of her leading literary critics. I also saw the recent movie versions of *Sense and Sensibility* and *Persuasion*, falling in love with Emma Thompson and Amanda Root in quick succession, and finished off my orgy with viewings of the BBC videos of *Sense and Sensibility*, *Emma*, *Mansfield Park*, *Northanger Abbey*, and *Pride and Prejudice*. Throughout—at least when I could remember to pay attention—I had two questions in mind. What does Jane Austen have to say about people, communities, and nature? And what is the cause of her resurgent popularity? Perhaps, I allowed myself to think, the questions are related.

Answering the questions proved not so simple, but I did have fun trying. Sam and I read Aunt Jane's letter, dated 8 Jan. 1817, to her nine-year-old niece Cassy, beginning:

> Ym raed Yssac
> I hsiw uoy a yppah wen raey. Ruoy xis snisuoc emac ereh yadretsey, dna dah hcae a eceip fo ekac . . .

I read the amusingly mordant comments she could write about her neighbors, such as the one in her letter of 3 July 1813 to her brother Francis, mentioning the "respectable, worthy, clever, agreable Mr Tho. Leigh, who has just closed a good life at the age of 79, & must have died the possesser of one of the finest Estates in England & of more worthless Nephews and Neices [*sic*] than any other private Man in the United Kingdoms."

I read the last chapters of *Pride and Prejudice, Emma,* and *Persuasion* each three times. I read once again about Catherine Morland's cruel expulsion from Northanger Abbey, and about the ill-omened trip of Fanny Price, the Bertram sisters, and the Crawfords to the Rushworth estate, Sotherton, with its seductive, if too regularly planted, wilderness. And again I was privileged to accompany Emma Woodhouse, Miss Bates, Frank Churchill, and Mr. Knightly on the tension-charged picnic to Box Hill, surely one of the highest peaks in English literature. Yet the answers to the questions remained elusive.

As for nature, Jane Austen has very little to say directly about it. Nature in Jane Austen's works is like nature in the Hebrew Bible; it is there as a constant presence, it is an essential fact of life, yet it is rarely separated out for special comment. Nearly all of her surviving letters—those not destroyed after her death by her sister Cassandra—are fully taken up with recording the details of caps and gowns; the observing of births, courtships, marriages, sicknesses, and deaths; and with commenting concisely and acutely on the characters of the people who filled her life. True, she is keenly aware of nature. For example, in her letter of 8 Nov. 1800 to Cassandra, she writes,

> We have had a dreadful storm of wind . . . which has done a great deal of mischeif among our trees. . . . I was sitting alone in the dining room, when an odd kind of crash startled me—in a moment afterwards it was repeated; I then went to the window, which I reached just in time to see the last of our two highly valued Elms descend into the Sweep!!!!! The other . . . taking a more easterly direction sunk amongst our screen of chesnuts and firs, knocking down one spruce fir, beating off the head of another, & stripping the two corner chesnuts of several branches in its fall. . . . And what I regret more than all the rest, is that all the three Elms which grew in Hall's meadow & gave such ornament to it, are gone.

But she quickly puts this loss in unromantic perspective. "I am happy to add however that no greater Evil than the Loss of Trees has been the consequence of the storm. . . . We greive therefore in some comfort."

In *Sense and Sensibility*, the materialistic, soulless John Dashwood describes to his sister Elinor, heroine of the book, how he is improving the estate that used to be her home. On the knoll behind the house, he is putting up a greenhouse. "The old walnut trees are all come down to make room for it." Elinor keeps her feelings to herself, but we know her sense of loss. Tony Tanner, one of Jane Austen's most perceptive critics, comments on this revealing scene.

> Jane Austen would not be the first person to feel that there are some trees better left standing, and some greenhouses better left unbuilt. But she was not sentimental about wildness and she recognized that society is necessarily a more-or-less continuous depredation of unchecked nature. What is implied in all her work is that human society ought to be very good indeed to justify the inroads made on "nature."

As Tanner observes, Jane Austen is less directly concerned with nature than with the human society that shapes and alters it. Even so, she is well aware of the critical connection between the integrity of communities and the integrity of nature.

If they are not odes to vanished landscapes, what then is the source of the popularity of her books in these uninspired and waning days of technological oppression and commercial empire? In trying to answer this question, I soon realized that I had made a mistake in undertaking the task. How am I—a person who never watches television, who seldom goes to public meetings, who rejects polls and the idea of polling, and who mistrusts pundits who speak knowingly of the public mind— supposed to figure out why America likes Jane Austen? No, better stick to safe ground. I will tell you why *I* like Jane Austen.

Having made my decision, I feel not only relieved but convinced of its appropriateness. For Jane Austen herself would never have claimed to understand the views of the British public. That is very much a twentieth- and twenty-first-century sort of presumption. She knew better than anyone else the intricate rules—namely, manners—that kept and still occasionally keep communities from unraveling, and she was as alive to the depths and subtleties of human character as anyone of her time or since; but she would not have believed that she knew what all England was thinking. On 1 April 1816 she put something like this sentiment in a letter to James Stanier Clarke, who had written to her to convey the acknowledgment and thanks of the Prince Regent for receipt of her novel *Emma*, which was dedicated to the prince. His Royal Highness had just appointed Clarke to be Chaplain and Private English Secretary to the Prince of Coburg, and, full of his own importance, he made the idiotic suggestion that Austen's next book should be a historical romance about the House of Coburg. Jane Austen's polite but firm reply follows:

> I am fully sensible that an historical romance, founded on the House of Saxe-Cobourg, might be much more to the purpose of profit or popularity than such pictures of domestic life in country villages as I deal in. But I could no more write a romance than an epic poem. I could not sit seriously down to write a serious romance under any other motive than to save my life; and if it were indispensable for me to keep it up and never relax into laughing at myself or at other people, I am sure I should be hung before I had finished the first chapter. No, I must keep to my own style and go on in my own way; and though I may never succeed again in that, I am convinced that I should totally fail in any other.

Jane Austen had two brothers who became admirals, one member of her family circle who was executed by the Committee of Public Safety in revolutionary France, and others who figured in sensational British trials, plus a working knowledge of city life. She was not a simple rustic confined by ignorance to write only about country villages. No doubt she could have expanded her artistic horizons had she chosen to do so. That she did not so choose is of course the foundation of her success and her great discovery. For her, the community was the fertile center of the world, and that was the ground in which she set her plow. She had learned and written about the types of human interaction that cause communities to flourish or to founder. She knew that any tales she might write of faraway romantic princedoms, of bloody revolutions, of naval battles and urban squalor, could only be less important and less profound than the "pictures of domestic life in country villages" that were her stock in trade.

When Jane Austen delineates the dynamics of small communities, she is at her best. She makes us see and feel that the deepest foundations of civilization are anchored in the little words and deeds that pass between friends and neighbors. The best example I can find occurs during the picnic at Box Hill, in *Emma*. Emma Woodhouse, confused by powerful emotions and social interactions which she is not yet mature enough to understand, makes a playfully cruel remark to the kindly but boring spinster, Miss Bates. Just a sentence, but so out of keeping with the thoughtfulness and sensitivity that have earned Emma a leading place in the community, that it shocks us like an explosion tearing across the beautiful landscape. Mr. Knightly, deeply in love with Emma, nevertheless finds it necessary to rebuke her.

> "I cannot see you acting wrong, without a remonstrance. How could you be so unfeeling to Miss Bates? How could you be so insolent in your wit to a woman of her character, age, and situation? . . . She felt your full meaning. She has talked of it since. I wish you could have heard how she talked of it—with what candour and generosity."
>
> "Oh!" cried Emma, "I know there is not a better creature in the world: but you must allow, that what is good and what is ridiculous are most unfortunately blended in her."

Mr. Knightly admits the truth of this, but points out to Emma that her knowledge of Miss Bates's poverty, age, and history made this careless remark inexcusable—by implication, the community's survival depends on each of its members paying attention in their words and deeds to the known personal status of every other member, a task requiring constant, if often unconscious effort. In her carriage on the return, "Emma felt the tears running down her cheeks almost all the way home, without being at any trouble to check them, extraordinary as they were."

I like Jane Austen because she reminds me, as Emma was reminded, of the things that ought to matter in my life, or, to put it differently, the things that would be allowed to matter in a world less dominated by superficial interactions, a world more bent on survival. It does me good to think about, even to imagine myself in, a society where there are rules of behavior that are clearly formulated and observable and that make sense. A society in which the moderate pace and human scale of life admit the possibility of my understanding what is going on around me. I like to think of a time when terrible consequences were commonly known to be firmly linked to violations of the rules that govern the relations between people in a community. In *Mansfield Park*, Sir Thomas Bertram reflects on the reason why his younger daughter is a disappointment and why his older daughter has disgraced herself and her family:

> He feared that principle, active principle, had been wanting; that they had never been properly taught to govern their inclinations and tempers by that sense of duty which can alone suffice. . . . Wretchedly did he feel, that with all the cost and care of an anxious and expensive education, he had brought up his daughters without their understanding their first duties, or his being acquainted with their character and temper.

I like Jane Austen because she describes a world in which people can make clear choices between honorable and dishonorable behavior, and can live to see that their choices matter. This is not to say that the choices in her books are all black and white. She knows, as Anne Elliot discovers in *Persuasion*, that there are "cases in which advice is good or bad only as the event decides." Nor does it mean that she thinks that dishonorable behavior always brings material punishment. Neither Wickham in *Pride and Prejudice*, nor Willoughby in *Sense and Sensibility*, nor Henry and Mary Crawford in *Mansfield Park*, nor William Elliot in *Persuasion* suffer materially from their reprehensible conduct. But they do suffer consequences, not least of which is that the community judges and rejects them when their serious flaws of character are perceived. This rejection then serves to restore the moral symmetry that the community has lost.

In a world such as ours, morally anesthetized, quick to rationalize and approve every self-destructive tendency, largely devoid of the living awareness of any higher purpose or power, I find it harder and harder to remember what community is, or the evolved systems of speech and action needed to keep community together. Jane Austen reminds me of what I have forgotten and tells me that which I never knew. Moreover, she shows me that it is possible and necessary for us to weave durable patterns of propriety into our own lives, patterns in harmony and balance with the patterns around us.

And with all this, Jane Austen, who died at the age of forty-one of a disease that would be described as treatable today, tells me that there can be constancy, love, and happy endings in the world, and that the currents of nature, the currents of community, and the currents of the soul can flow together for a while in a blissful bubbling stream that might tempt God to smile.

UNIVERSITIES AND THEIR COMMUNITIES

I received an invitation to speak to the Heinz Endowments a few years ago. This major foundation was thinking of starting a program of charitable donations to help the environment and wanted advice about how to make the best use of its money. Would I participate, the director asked, in a one-day meeting on environmental education being organized by David Orr? My topic would be the role of the university. I went, and the following is more or less what I said to Heinz.

During my first years as a board member of the Educational Foundation of America, which gives grants in a number of areas, including the environment and education, I was struck by the extreme scarcity of exciting, innovative, *useful* proposals coming out of the major research universities: Harvard, Princeton, Stanford, Berkeley, and the like. The second-tier research universities are no better; they are all scrambling to copy the bad models ahead of them.

The problems that the universities are doing little or nothing to address — either in teaching or in research — are those that we must confront if our civilization is to survive. They are materialism in our culture; the deterioration of human communities; anomie; the commercialization (privatization) of former communal functions such as health, charity, and communication; the growth imperative; exploitation of the Third World; the disintegration of agriculture; our ignorance of the ecology of disease, especially epidemic disease; the loss of important skills and knowledge; the devastating decline in the moral and cultural-intellectual education of children; the impoverishment and devaluation of language; and the turning away from environmental and human realities in favor of thin, life-sucking electronic substitutes. Far from confronting these problems, universities are increasingly allying themselves with the multinational commercial forces that are causing them. The institutions that are supposed to be generating the ideas that nourish and sustain society have abandoned this function in their quest for cash.

It is typical, for example, that with all the academics working on developing and patenting new crops, the only effective mechanism for monitoring and preserving the priceless and rapidly dwindling stock of *existing* crops in North America and Europe—the heritage of agriculture—was developed by a young farmer completely outside the university system. Kent Whealy's dream was originally funded by a $6,000 foundation grant; now his Iowa-based Seed Saver's Exchange helps gardeners and farmers all over North America and much of Europe exchange and grow thousands of varieties of fruits and vegetables that would otherwise have been lost because they are no longer available from commercial seed companies.

Any one of the interrelated problems I've listed could destroy Western society. So why are the universities, which society counts on for answers to its most pressing problems, doing next to nothing? It's easier to get academics at, for example, Yale and Columbia to think about processes and inventions that will bring in cash and prizes, or about abstract problems in faraway places, than it is to get them to think deeply and constructively about why New Haven and New York are so physically and socially distressed and why their students run the risk of being assaulted on the way to class. Or why, as Jonathan Kozol has told so powerfully, only a few miles from university hospitals where the rich can receive every attention, there are hospitals where desperately sick, poor Hispanics and African Americans must spend days in corridors waiting for a room with a bed—and when they finally get one must change the bloody sheets themselves before lying down. There are, of course, exceptions. Muhammad Yunus, former head of the Economics Department at Chittagong University in Bangladesh quotes, in his book *Banker to the Poor,* a blistering attack from a friend:

You academics are failing us. You are failing in your social duties.
. . . No one is accountable to anyone for anything. Certainly not
you lily-white-handed academics with your cushy jobs and your
jaunts abroad. You are useless all of you. Utterly useless! I am absolutely disgusted by what I see in this society. No one thinks of the
poor.

Yunus turned his active concern for the poorest people in the villages
around his university into a new system of making small loans within a
context of social reinforcement—microlending—which would lift the
poor, especially women, out of perpetual debt and exploitation into
a position of working independence. Within a few years, Yunus's
Grameen Bank had become a major force for social justice throughout
Bangladesh, and its ideas had been applied in many other countries, including the United States. But examples like this are vanishingly rare.

Even many ecologists, who ought to understand the importance of
community interactions as a factor in survival, seem disconnected from
their communities. In a paper submitted for my graduate course in conservation ecology, a student, Aviva, cited the famous statement of Rabbi
Shimon the Righteous (or Shimon the Wise) in the Jewish "Ethics of the
Fathers." Shimon said, "The world is sustained by three things: Torah
[study of the law], labor [implying also reverence to God], and acts of
loving kindness," the last of which the student translated as "community
service." Her paper was an explanation and description of her unsuc-
cessful efforts to introduce the idea of adding community service to the
requirements for a graduate degree in ecology at Rutgers, a daring suggestion in today's academic world.

Aviva's education before coming to Rutgers, the state university of
New Jersey, was in Jewish religious schools, from kindergarten through
college. At some of these kinds of schools, including hers, a code of ethical behavior that connects the individual to the community is considered a normal and necessary part of a student's training. Although religiously mandated, this teaching is not separated from the secular
curriculum; when the opportunity arises to combine the study of ethical
(and other) religious laws with mathematics, biology, or literature, it is
done. Therefore, when Aviva perceived a potentially useful connection
between ecology and community service, she was simply using her prior

education. Her experience did not prepare her for the self-centered academic climate of most secular universities, including Rutgers. True, all universities have service activities that their students can participate in, but they are generally walled off from and considered secondary to the academic curriculum. The academic curriculum itself reflects the research priorities of the specialized disciplines: if the public is part of these priorities, it is the generic, abstract public—not the folks right outside the gates.

All major religions make ethical behavior and community service a respected part of education. These activities also used to be a function of the universities and land grant colleges, although there were signs of decline as long ago as the first decades of the twentieth century. Today they have become almost vestigial, like the human appendix, the province largely of aging professors, a few younger, tenured mavericks, and those assistant professors who are most likely to be terminated when their seven years are up.

Universities have lost their ethical direction and sense of responsibility to the community for at least four reasons: bloated, top-down administration with its own corporate rather than academic and communal priorities; professors who are selected and promoted for ambition and ruthless opportunism, which go along with a lack of commitment to the encumbering and time-wasting ideal of community service; the specialization that brings in grants even as it obliterates all hope of seeing and responding to the big picture; and the fact that many universities are too big and unwieldy to meet any kind of serious challenge effectively. In other words, universities mirror all too well the rest of modern culture. Under these circumstances, a foundation grant or gift to a modern university has the potential of doing more harm than good.

But you here at Heinz, I said to them, have money that the IRS says you had better give away—how should you spend it? First, identify the problems you want to concentrate on and forget the universities unless they are going after the same problems. My own starting point and prime concern would be the children. If we don't do something on a grand scale and do it quickly, the future adults who are now four, five, six, and seven years old are going to be incapable of saving their country when they grow up. They need while they are young the real experience of nature, the real experience of group (community) learning and cooperation, and the real experience of the joy of sustained and satisfying work. None of these things can be done by watching television, by surfing the 'net, or by absorbing the competitive values of a celebrity- and money-centered society.

This is what you might do. Help make kindergarten to twelfth-grade teaching an attractive profession—capable of getting and holding the cream of the crop of graduates from our best colleges, many of whom would like to teach if they thought it were a viable career. Develop ways of subsidizing the teachers of environmental studies, especially ecology and conservation, so that schools will not consider them expendable. Make sure that they have supplies, money for field trips, money for their own training, and a secure income that will not stop when the grant stops. Reward innovative, successful teachers. A so-called developed world that gives Nobel prizes to selected economists and bare-bones salaries to master kindergarten teachers is not going to last much longer. Support university projects that get involved with children, like the Rainforest Collaborative of the University of Massachusetts/Boston, which took children from inner-city Boston schools to Massachusetts forests to learn ecological principles in the field, and then took selected children from this group, with their teachers, to Costa Rica, where they discussed the ecology of very different kinds of forests with local children. But don't support such projects unless they intend to stick with the children, interact with their parents, stay in touch with their teachers, follow the students through high school, and guide them into college.

Here are some more suggestions: Give to locally oriented projects rather than ones aimed somewhere else. Give to small institutions rather than to big ones. If you give to big institutions, don't pay any overhead unless it is earmarked for direct project support rather than an administrative rakeoff, and don't pay anything unless the university chips in real dollars, too. I don't mean the phony "contributions" of space and salary that would be provided anyway.

Remember that you can't do philanthropy from the office. I use the excellent example of the Educational Foundation of America, in which every solid proposal gets at least one site visit either by a trustee or by the executive director or one of her assistants. Beware of proposals that were written entirely in institutional development offices. They are easy to spot because they are all alike—the same superlatives, the same naïvely inflated claims, the same lack of passion and originality. You should seek out the small organizations, often just two or three people, that are doing exciting work, and help them develop proposals. If you find some of these people, you'll find others; there is a network out there.

Favor integrated proposals that involve institutions and their local communities working together, preferably including both children and adults — and make sure that they have a good chance of being self-sustaining after you end your support. Don't support academic conferences, but do figure out ways of helping your grantees spread the news about successful projects so that others don't have to reinvent the wheel. Develop strategies that force universities to spend their money on the right things; do not give them environmental funding if your grant is just going to free up other money to hire more genetic engineers. It is essential that university adminstrations be cajoled, embarrassed, threatened, and, if necessary, starved into changing the reward structure so that creative, energetic professors who are more interested in teaching, community service, and important but unglamorous investigations than in big-dollar research are honored and promoted.

Finally, beware of us experts — too often we are people who are extremely knowledgeable about narrow systems that don't apply to the real world and never will, systems that bear no useful, time-tested relationship to the lives we are living. The great University of Wisconsin botanist and conservationist Hugh Iltis told me the following story about experts at the international corn and wheat research and breeding institute, CIMMYT, in Mexico. It was 1981; Iltis, hot and sticky from a field trip, walked into the faculty-student bar at CIMMYT for a beer. He was still carrying a bundle of annual teosinte, a Mexican grass generally believed by taxonomists to be closely related to the ancestor of corn (maize). Joining a group of advanced students and professors at a table, he threw down the teosinte, which looks nothing like corn, and asked, "What is it?" Eighteen or twenty experts gathered around, examining and muttering. Nobody knew. After a while, the waitress, a short, black-haired young woman of obvious native descent, arrived carrying a tray of beer glasses. "What is it?" Iltis asked her in Spanish, pointing to the teosinte. Glancing at the grass briefly as she set down the glasses, she answered, *"Madre de maíz"* — the mother of corn.

Our American empire is dying, along with the empires affiliated with it. The universities and other institutions we look to for guidance are dying, too; they have become dull, mean-spirited, greedy, and tired. In the past, dying empires have not recovered—they seem to gather a fatal momentum of decline. But there are still in America and the rest of the First World vast resources of wisdom, creativity, and philanthropy, as well as the energy to use them. I see at least a half-dozen students like Aviva every year, each worth his or her weight in diamonds. And I see, outside the universities, hundreds of mostly small, community-oriented organizations that could use these students' imagination and spirit if there were money available to pay them a living wage. Here is where you and other like-minded foundations can help. It will take a miracle for us to survive our crisis of spiritual and environmental health, yet miracles sometimes come to those who work and wait for them. What other choice do we have?

AN INVALID'S GUIDE

I have had some more strange dreams—two of them, a day apart—but not so somber as the one about the vine cleaners. They occurred in the early morning just before I awakened, which is why I remember them so vividly. In the first dream, on a Wednesday, an insurance salesman is trying to sell me life insurance.

"You need it, Marvin," he argues, "to protect your wife and children in case, God forbid, something happens to you." I am not convinced. I have plenty of insurance, I think to myself. And the salesman's manner irritates me—I dislike it when people I don't know call me by my first name. Besides, my name isn't Marvin; it never has been. Is it more polite for him to call me Marvin rather than David, I wonder? Before I can respond to the salesman, I wake up and the dream ends.

What does it mean, I ask myself? Certainly there is no ominous feeling about the dream—in fact it leaves me laughing. And why Marvin? A pretty pass things have come to, when a person's subconscious forgets his first name. Could I have gotten into someone else's dream? Hardly likely, I reason. For a little while I puzzle over the dream's significance with no results; by the time I am finished shaving, I have put it out of mind.

Thursday's dream brought more light. In this one, I have just written a book—or maybe it's a pamphlet. Now comes the hard part; I have to find a title for my work. I am particularly bad at coming up with titles, a fact that I am well aware of even when I am not sleeping. (My best title, *The Arrogance of Humanism*, was actually the brainchild of my former editor at Oxford University Press, James Raimes.) But in the dream I am inspired; a brilliant title flashes into my mind like a comet. I will call my book (or pamphlet) "An Invalid's Guide to Putting Out Mine Fires."

No matter whether it is directly relevant to what I have written—the end of the dream is fast approaching and I am no longer sure—who cares? The title is enormously pleasing: witty, catchy rhythm, full of some sort of meaning. As the dream fades into slow wakefulness, I remain pleased with myself and in excellent humor. The absurdity of my title tickles me. Two days in a row of feel-good dreams, a record for me. What's going on?

I stay in bed trying to figure out the two dreams. I am convinced that they are related, but how? The connection eludes me. There does not seem to be an obvious relationship between my insurance salesman and my book title, regardless of how I twist them about. Then, when I have momentarily let my thoughts wander, it comes to me.

"It's a joke, stupid," an inner voice whispers. "They're both jokes. Lighten up."

I feel a wave of gratitude and relief. Think of it! After more than fifty years of specializing in nightmares and crude horror, my subconscious has started sending comedy down the pike. Good old subconscious! I thought it had it in for me, but it really cares.

Encouraged by this insight, I begin to examine the dreams more closely. Humor aside, they must mean something, I figure. They must relate to something in my life. What preoccupied me in the days before the dreams? That was an easy question to answer: I had been brooding about Bill Clinton's and Al Gore's Republican-style enthusiasm for NAFTA, the North American Free Trade Agreement, and the WTO, the World Trade Organization, and their deafness to those of us who had concerns about the environmental, economic, and social consequences of "free trade" controlled by multinational corporations. About

their affinity for the likes of Tyson Foods and Monsanto, and their alienation—in practice if not in speech—from the nation's small farmers and
farm communities. And I had been wondering why my wife and I had
given the first Clinton-Gore campaign our support and more money
than we care to remember, only to find ourselves saddled with politicians
whose beliefs and principles change as rapidly and easily as the colors of
a chameleon, always quick to match the dominant shades of their surroundings.

Still in bed, I muse about this and slowly begin to see what the first
dream means. My streetwise subconscious, in its new, avuncular role, is
giving me a much-needed lesson.

"You won't buy life insurance from a salesman who calls you Marvin,
but you gave Clinton and Gore all that dough. I always knew you had
the brain of a chickpea, but I expected more from your wife. You saw
that Clinton was slippery as an eel during his first campaign—grinning
and pointing at Gore any time anyone asked him hard questions about
the environment—but you heard what you wanted to hear. Now you're
really stuck; with the ruthless disciples of Gingrich like DeLay and
Armey, with the unspeakable Lott, and with George W.'s crowd preach-
ing "bipartisanship" but behaving like marauding warlords on behalf of
big oil, nuclear power, timber, agribusiness, and mining—all of them on
one side, and with a pack of wimpy, semi-principled Democrats also ob-
ligated to big business on the other. Where are you gonna go?"

When politicians get to your subconscious, you know times are bad.
Unwilling to get out of bed, I think about all the other things that were
on my mind in the days before the dreams. The new "kindlier conservative" prescription for the environment includes the following: legislation to make it easier to build roads in national parks and wilderness
areas; to lower standards for drinking water and impose a moratorium
on the Clean Water Act; to prohibit new listings of endangered species
and to cripple or kill the agency that studies them; to cut funds for ocean
fisheries protection and for research to help save the Pacific salmon; to
allow salvage logging that will devastate what's left of our Federal forests;
to make sure that the century-old laws governing mining on public lands
remain anti-public and pro-corporation; to make it much easier to develop wetlands; to drill for oil in the Arctic Wildlife Refuge, seriously

damaging the refuge for a possible six-month supply; to weaken the EPA; to protect the factory farming of hogs and chickens that is harming entire counties in the southeast and midwest; to reduce public access to information about pollution from factories; to further the spread of genetically modified foods and hamstring the massive public opposition to it; and—just in case we didn't have enough to worry about—to reduce or eliminate screening for deadly pathogens in our meat supply. All of these profound changes are to be enacted with hardly any public debate.

When you read this chapter, most or all of the items I mentioned will have been voted up or down, as the case may be. But the ones voted down will return and there will be new challenges to our national decency and survival, because the goals and policies of the multinational lobby, which controls both political parties, will remain the same as those I was brooding over while lying in bed on that Thursday morning. Their effect will be to keep us permanently off balance with multiple assaults coming from different directions, each more outrageous than the last. To make us feel weak and helpless, like . . . like an invalid trying to put out a mine fire, which breaks out first in one inaccessible place and then another.

There it was, the explanation of my second dream. You knew it was coming, but I didn't. What an image of hopelessness! How can we face such overwhelming power, such dreadful odds? But why had I found my dream so damned funny? Is the ruining of the world a joking matter?

I get out of bed and quickly shower and dress. I am disgusted at the insensitive behavior of my subconscious—laughing while everything around me burns. But on the way to the office, my subconscious makes its case. A familiar voice whispers, "Think about Natan Sharansky."

"What about him?" I respond tersely.

"He was a refusenik in the former Soviet Union, persecuted for practicing and teaching his religion. The government sent him to Siberia and abused him horribly for years—he was cut off from his wife, his friends, and his work, half-starved, sick, subject to the cold of the Siberian winter and the heat of summer, tormented by his captors. . . ."

"Yes, I know about all that. Get to the point."

"Well, what happened while they were finally letting him go, in 1986?"

"They took him to the old East German–West German border and released him. What does that prove?"

"Don't you remember what happened when they got to the East Berlin airport?" the voice replies with a touch of impatience. "As he deplaned they ordered him to go to a waiting car. Freedom was a few miles away. The world would be expecting him on the other side of the border. 'Go directly and quickly to the car!' the Soviet officials barked at him, noticing the propaganda ministry's cameramen waiting to film his movements. And what did Sharansky do? As the movie cameras rolled he zigged and he zagged. He walked sideways and backward, maddeningly slowly, toward the car and away from it, while the Soviets watched in helpless rage. Now do you understand?"

"No," I answer, losing patience in turn. "He remained a refusenik until the very end. How is that relevant?"

"Listen!" the voice yells. "Will you listen? He did something more important than that. While he was still in their control, under their yoke, he laughed at them and mocked their power. Sharansky, who knew how grim and terrible they were, made fun of them. Sure, this is a far cry from the Soviet Union, but you had better do some laughing, too."

"All right, I see what you mean," I admit grudgingly. The truth is, my subconscious has grasped what my conscious mind has been slow to comprehend: as long as our country is run by hacks, bullies, sell-outs, and ignoramuses, the best way to keep ourselves sane and capable of effective action is to laugh at them loudly, publicly, and often. Laughter is surprisingly powerful, especially when you least feel like laughing.

As usual, my subconscious has the last word. "There's another reason to keep on laughing," says the voice. "I'll tell you because I know you won't figure it out for yourself. It distinguishes you from most of the extremists and fanatics wandering around loose these days."

"How's that?" I ask.

"They never laugh," the voice replies.

SWIMMING LESSONS

The father is bound with respect to his son, to circumcise him, redeem him, teach him Torah, take a wife for him, and teach him a craft. Some say to teach him to swim too. R[abbi] Judah said: He who does not teach his son a craft teaches him brigandage. . . . What is the reason [to teach him to swim]? His life may depend on it.

The Babylonian Talmud,

tractate Kiddushin, 29a, 30b

Shortly before the birth of our first child, Kate, now twenty-eight, Joan and I bought a copy of Dr. Benjamin Spock's *Baby and Child Care*. It proved to be very reassuring. A thick book with no wasted words, it told almost everything we needed to know about the coughs, the sneezes, the spots, the fevers, the crying at three in the morning, the rejection of nourishment, the emergence or nonemergence of teeth, and the terror-inspiring inclination of sleeping infants to stop breathing for a while just when their parents are getting settled in bed.

There were, of course, a few things Dr. Spock failed to mention. I knew from medical school the signs and symptoms of many rare and terrible diseases of infants. Kate got them all. Each time we would rush her to the pediatrician, Dr. Arky, and each time we had the satisfaction of bringing a little uninhibited laughter into the life of that overworked and kindly man.

Fortunately for Dr. Arky, we moved from north to central Jersey when Kate was two and found a new pediatrician, Dr. Lapkin. Although we got more experienced as time passed, and came to know Dr. Spock pretty much by heart, there was always something new. I remember, for example, my deep concern about Jonathan, our third child, when he was an infant; he did not seem to be able to hear. Joan was not particularly worried at first, but fear is contagious. What if the development of his language were affected? Anyway, it was time for his one-month visit to the pediatrician.

Dr. Lapkin appeared somewhat tired but was in a good mood; it was nearing the end of a long day. I watched as he listened to Jonathan's heart and lungs with his stethoscope and examined his reflexes.

"He looks fine," Dr. Lapkin said. "Any problems?"

"He doesn't seem to hear loud noises," I said. "I'm afraid that he may be deaf. Is there some kind of audiometer test you could give him?"

"He's responding normally to my voice as I speak," answered Dr. Lapkin. "He has been during the whole examination. Are there any other problems?" he asked, directing the question to Joan. As he left the examining room he was laughing—his laugh sounded a lot like Dr. Arky's.

Jonathan, now seventeen, actually does have some deafness, a curious selective deafness that keeps him from hearing us when we call upstairs for him to get off the phone and come wash up for dinner. His sisters and brother all have it too. It must be hereditary.

Not all visits to the pediatrician end with good news, but most do. My point is that at least for a child's early years Dr. Spock or its equivalent, plus a good pediatrician, will do fine for nearly all of the crises that a parent will face. As children grow older, however, no book, no matter how thick and concise, and no pediatrician, no matter how competent, can possibly be expected to answer every question that arises and solve every problem that comes up. Children do things that are not described in any books, unique things that you the parent must learn to cope with by yourself, without the help of paid experts.

To start with a simple example, there was the day when Kate, age five, came home from kindergarten walking oddly. She was dropped off by the carpool at the end of our very long, gravel driveway; as we watched her from the doorway coming toward us, she oscillated from side to side, singing happily. When she got closer, we were relieved to see that the cause was not medical—she was wearing only one shoe.

"Where's the other shoe?" Joan asked as she entered the house. Kate looked down at her feet, surprised.

"I don't know," she answered in a disinterested tone, "maybe in the car." Then she hobbled off to the kitchen to get a drink of orange juice.

And there was Jane, not quite four years younger than Kate. During her early years we couldn't figure out her strange lack of interest in possessing material objects. Kate's room was filled to the Plimsoll mark with bits of colored glass, pebbles, stuffed animals, seashells, wilted flowers, china figurines, and other forms of solid waste—Jane's was as empty as the Gobi desert. That empty room worried us (although Jane was not concerned). Were we neglecting her? Was she normal? On her third or fourth birthday—I've forgotten which—we made a supreme effort to overcome Jane's antimaterialism and presented her with a personable and charismatic stuffed elephant almost as large as she was. Jane seemed pleased, if not actually overcome with joy. There was a barely suppressed gleam in Kate's eye; she said nothing. A few days later the elephant went to live in Kate's room. We offered to escort it back, but Jane declined the offer; she had tried materialism and found it wanting.

Enjoyable as it is to record these anecdotes—some minor problems faced by parents of good children living under affluent conditions in a quiet community—it is time to get down to business and confront the hard questions about being a parent. As children and parents age together, the questions become ever more difficult to answer. Eventually, one comprehensive question emerges, often too late for the answer to be of use: What are the responsibilities of parents in these days of social and environmental disruption and turmoil?

This is hardly a new question; every religion and culture has grappled with it. Judaism, my own tradition, is exceptionally child-oriented, preoccupied with the survival of the people through its posterity and with the effects of the behavior of parents on the children. The Talmud, the principal storehouse of rabbinic wisdom about all phases of life, a vast work of many volumes, is the obvious first place to look for a detailed compendium of the obligations of a Jewish parent. After all, the Talmudic rabbis even speculated about an arcane subject like artificial insemination in humans, which was then more than a millennium and a half from being developed. But if we search for a Talmudic equivalent of Dr. Spock, an encyclopedic list of responsibilities and operating procedures for parents, we will be disappointed. I have quoted at the beginning of this chapter the nub of the most celebrated statement about parental responsibilities in the Babylonian Talmud. If you look back at it, you will see that it is, like the original U.S. Bill of Rights, surprisingly short. In rabbinic writings, when a subject of this importance is dealt with concisely we can take it as an invitation to look beneath the surface meaning of the words.

Before beginning, however, I had better deal with the issue of sexism. The Babylonian Talmud is based on three centuries of rabbinic discussions that were begun approximately eighteen hundred years ago. Although enlightened for its time about the role and value of women, it falls far short of enlightenment by secular, twenty-first century, industrial world standards. For example, a father's obligations to a daughter, discussed elsewhere in the Talmud, concern only her physical need for clothing and support, plus the provision of a generous dowry so that she can marry advantageously. Fathers and mothers, sons and daughters, are treated very differently. Fortunately, there is a way around this modern difficulty that does not force us to jettison the enormously valuable Talmudic teachings. Without pretending that this was the intent of the Talmudic sages, I will simply substitute "parent" for "father" and "child" for "son" in my own consideration of the Talmudic passages. Thus I can explore the timeless wisdom of the Talmud's advice on raising children without being sidetracked by other issues.

Let's get back to the quote. Circumcision at eight days and the redemption of the first-born son pertain only to male Jews—they are not part of an ecumenical scheme of advice to all parents, and I won't discuss them further. The same could be said of the obligation to teach Torah, which in its narrower meaning is the first five books of the Hebrew Bible, but which in this case means the entire structure of Jewish law and commentary, including the Talmud. But I think that this injunction has a value even for non-Jews. It says that a parent is obliged to see that a child gets detailed, practical, ethical, consistent instructions for relating to God (or whatever higher power one wishes to substitute), the community, and the world around.

The injunction to teach Torah to one's children is underscored in the Jewish prayer known as the Sh'ma. The Sh'ma, whose name comes from the first Hebrew word of the prayer, "Listen!" is the central prayer of Jewish worship—the first prayer taught by a mother (or father) to a Jewish child and the last prayer recited by a dying person. It is spoken at least twice a day, morning and evening, by ritually observant Jews. The first paragraphs of the Sh'ma, taken from books 6 and 11 of Deuteronomy, contain clear instructions regarding children: "And the words which I command you this day shall be upon your heart; and you shall teach them diligently to your children, talking of them when you sit in your house, and when you walk by the way, and when you lie down, and when you rise up." In other words, education about correct religious and ethical behavior is the responsibility of the parent—it can't be delegated —and it takes time, a significant amount of the available time during waking hours. I think this obligation can be generalized to parents of any religion, and to nonreligious parents as well. In the latter case, the ethical system becomes paramount.

Some of the lessons of Torah are hard to teach and hard to learn. I'll give an example. Among the 613 biblical commandments that Jews recognize (of which some have not been observed since the destruction of the Second Temple in the year 70 C.E.), are several commandments that prohibit gossip about others. This applies to any idle talk about third parties, including favorable talk, for even friendly gossip about others can be an indirect cause of damage—for example, by provoking jealousy. But gossip is a central part of the lives of school-aged children, and they come home from school bursting with it. It is not easy to explain to a young child that it is all right to say "Billy hit me during recess," because you the parent should offer reassurance and may have to do something about it, but it is not all right to say "Billy hit Joey and Michelle during recess." As always, a parental example is the best teacher, although this is easier in the writing than in the doing. Still, the lesson can be learned. When our youngest, Sam, announced at the dinner table one night when he was ten, "One of the boys in the class argued with the teacher for five minutes about his grade, and then he threw a book at her and got sent to the office," he was aware that Joan and I really wanted to know the boy's name, but that we wouldn't ask. And we knew that he wanted to tell us who it was, but would resist the impulse. The shared feeling that comes from this kind of mutual restraint is very gratifying.

What happens if parents do or don't teach their children Torah? The answer is given in the second paragraph of the Sh'ma:

> If you hearken diligently to My commandments . . . I will give the rain of your land in its season, the former rain and the latter rain, that you may gather in your grain, and your wine, and your oil.

And I will give grass in your fields for your cattle, and you shall eat and be satisfied. Take heed to yourselves, lest your heart be deceived, and you turn aside and serve other gods, and worship them; and the anger of the Lord be kindled against you, and He shut up the heaven, so that there shall be no rain, and the ground shall not yield her fruit; and you perish quickly from off the good land which the Lord gives you. Therefore you shall lay up these My words in your heart and in your soul. . . . And you shall teach them to your children, talking of them when you sit in your house, and when you walk by the way, and when you lie down, and when you rise up.

The question of whether an angry God withholds rain from people who violate the commandments and who don't teach them to their children is of limited interest here. The threat may be real or it may be a metaphor; no matter. There is, however, a message in these words of the prayer that I cannot ignore, but it does not come through in English translation. One of the troubles with English is that the second-person pronoun *you* and the possessive forms *your* and *yours* are the same in the singular and the plural. This may be why the slang word *youse* and the southern regional term *you-all* or *y'all* came into being in American English; they can fill our need for a distinct second person plural. In Hebrew, however, the singular and plural forms of *you* are separate and distinct; there is no ambiguity.

In the first paragraph of the Sh'ma—the injunction to love God and teach your children God's ethical and religious laws—the *you* is singular: each parent has an obligation to teach his or her own children. In the second paragraph, which deals with rewards and punishments, there is a striking change. The *you* who receives the rain or against whom the heavens are shut is suddenly a plural. In other words, if you and you and you and the majority of other individuals who make up the community live right and teach your children to live right, then the *community* reaps the reward. If you live badly and neglect your children's spiritual and ethical education, then it is the *community* that suffers and dies.

The rabbis who first created the Sh'ma from separate passages in the Hebrew Bible were farmers or the children of farmers. They knew, as good farmers know today, that the individual cannot long survive if the community dies, but the community itself gains its strength from its individual members who live just and righteous lives and pass this way of life on to their children. As Wendell Berry wrote in *The Unsettling of America*, "One must begin in one's own life the private solutions that can only *in turn* become public solutions. . . . The use of the world is finally a personal matter, and the world can be preserved in health only by the forbearance and care of a multitude of persons." These "private solutions" start with the teaching of children. As your teaching prospers, so does the world.

It is tempting to dismiss the Talmud's direction to find a wife for your son as an archaic practice that is in fact ignored by the great majority of today's Jews, including many religious Jews. But there is a modern interpretation of the teaching that bears study. We may not want or be permitted to go out and find spouses for our children, yet we can and should influence the process. We can try to define by our example and by our words what a good spouse is like; we can tell our children about the common early warning signs of trouble in potential mates. Our words may fall on deaf ears or they may arouse anger, but there is no telling when words—and examples—will be recalled at a time of need and serve their intended purpose; so the teaching cannot be neglected.

The instruction to teach your child a craft seems like the simplest item on the Talmudic rabbis' list of parental obligations. In pre-industrial societies, few people survived without a craft or trade; one had to able to *do* something. The unacceptable alternatives were starvation or, as Rabbi Judah suggested, banditry. Today in industrial countries few parents actually teach their children a trade. There are various reasons for this: children don't expect to take over their parents' businesses, vocational or professional training has become the province of specialists, technological systems are evolving so rapidly that parents cannot keep up, and many children postpone making decisions about jobs until after they have left home. Also, the nature of work has changed radically. In the majority of jobs, work involves the routine application of already defined systems—there is no room for craft. Such jobs are either self-teaching or require only a low level of training and experience. Is the advice to teach your child a trade obsolete?

My answer is no. If anything, teaching a craft to children is more important than ever; but "teaching a craft" has to be defined in contemporary terms. Except for a few, it no longer means learning the family trade from a parent. Coopers, Smiths, Wheelwrights, Bakers, Drapers, and Fletchers may carry on the family name, but that is all they carry. What a parent now must teach a child about craft is altogether different from how to hold a certain tool, how to find particular materials, or how to recognize certain natural signs and signals. (It is important in children's development, however, for them to have experience using tools, handling different materials, and observing at least some of nature's useful indicators.) No longer passed from parent to child are the little secrets that distinguish _our_ barrel, _our_ violin, _our_ way of growing rice, from those of all others. This is sad, but there is no time for nostalgia; we have new things to teach our children about work.

First, we have to teach children that they must have a craft, even if we do not teach it ourselves. Craft, in my expanded definition, is the special skill, acquired only through training and experience, to produce particular goods and services. More than that, craft is an attitude, a way of thinking about work, that directs our energies toward the goals of continuous self-improvement, careful workmanship, pride, and completion, no matter what the task. Also implicit in the Talmudic instruction is that one's craft should make the world a better place and that it should be informed by the laws that govern our relationship with society and the environment; otherwise, brigandage could be a craft. It is said that in the nineteenth century, Jewish teamsters in the Lithuanian city of Vilna, then a great center of Jewish learning, would suspend their physical labors at noon each day for an hour of intensive study of Torah. This is craft.

Many parents have a craft; children should see their parents at work and participate when possible. Joan is a fine example: a forest ecologist, her research is done both in the field and in the laboratory, and all of our children have watched and helped her at work at one time or another. They have had ample opportunity to observe her painstaking care, demanding standards, tireless persistence, her love of what she does, and the affection and respect her students give her. They have often seen her leaving for the Pinelands or North Jersey in the morning in all weathers, dressed in her field clothes and carrying her field equipment, and they are aware that she settles down at the computer in the evening—after their bedtime when they were little. They have noticed that unless she is helping them with their homework or reading one of my manuscripts, she takes time off only for a quick mug of tea and a cookie at 9:00 P.M. None of our children have followed or are headed towards Joan's particular career path, but there is no question that all of them have learned craft from their mother.

Mothers and fathers, however, are not the only teachers. The endless message of television, radio, Internet ads, school, friends, and neighbors to our children is that they are valued primarily as consumers. Our message has to counter this. "Become a producer" is the theme the children should hear from us; "your worth is determined by what you are and what you do, not by what you wear or consume." Despite its overwhelming prevalence, consumerism is vulnerable. We can fight it by good example and by ridicule. Totally passive consumers are feeble, clueless, exploited creatures, easily caricatured in a way that children understand. Consumers are weak. Children don't want to be weak. Even if we don't teach our children a particular trade, we can teach the joy of knowing how to do useful things, how to finish what you start, how to begin to take care of your own needs, how to do things right. How to produce.

We have to teach children that a job is not necessarily a craft. Repetitive, mindless jobs producing uniform products; jobs producing shoddy, unnecessary goods; socially destructive jobs; those bureaucratic or managerial jobs that add little or nothing to real production; servile or degrading jobs; completely redundant jobs; and jobs in which one's experience, talents, and contributions are ignored are not crafts. Although a child may have to hold one of the less offensive of these jobs at some time in his or her life, it will still be essential to have a craft and stay in practice. We must also point out that physical labor and the work of one's hands are not inferior to intellectual work. Many of the best crafts, from surgery to teaching to farming, can combine the two.

Increasingly, schools and especially colleges are training our children for jobs that will not exist or will not be worth having after they graduate. They are being educated for the supposed benefit of "the economy," not for themselves or their communities. They are schooled in technologies and trained on equipment that will be obsolete before they get their diplomas, or they are directed toward powerful, high-tech industries that want to hire fewer, not more, employees in the years to come.

An example is the communications industry: according to the Bureau of Labor Statistics, at the start of the communications revolution, between 1981 and 1988, telephone communications companies in the United States eliminated nearly 180,000 jobs, and the trend has continued unabated since then. A result of this kind of education and work experience is that it gets harder and harder to hold a job for more than a few years —a circumstance that makes craft impossible. If young children are taught the differences between craft and mere jobs, if they are taught to recognize and value the kinds of work that produce goods and services people truly need, then they have a better chance of finding satisfying, durable, productive work, even creating their own jobs, after they leave home.

Finally we get to swimming, my favorite part of the Talmud's list of instructions to parents. Because it is such a short list, the inclusion of swimming amuses me. Why swimming, of all things, in the Middle East where water is scarce and where, in ancient times, many people probably went from one year to the next without having an opportunity, let alone a need, to swim? If the Talmud had been written in Florida rather than Babylon, concern about swimming would have seemed more appropriate. So this instruction is a puzzle, and in the Talmud a puzzle is an invitation to pay attention and to think.

What the Talmud is saying to a desert-dwelling people when it states that they are required to teach their children to swim is profound and important. The meaning, I believe, is this: Your children may find themselves in various kinds of environments during their lifetimes. Some of these will be quite different from the places in which they grew up. It is your paramount responsibility as a parent to teach them how to survive in different environments that they may not be able (or want) to escape or avoid. Swimming is the ultimate example of such a survival skill. (The Talmud often uses extreme examples to make a point; it is understood that less extreme cases are included in the lesson.) In water, many of the things that come naturally to us as terrestrial animals, such as walking, running, resting, breathing, eating, seeing, and hearing, will not work in the usual fashion and have to be relearned. Children relearn much more easily than adults; you must teach them to swim early.

Although the need to learn to swim is real and the Talmud's instruction can be taken at face value, swimming also serves as a metaphor. The

Talmudic sages were clearly making a more general point, a point expressed as concisely as a thought in poetry. Make your children aware of the natural world around them—its many forms and manifestations, its moods and seasons, its creatures, its processes, its changes, its patterns, its predictabilities, and its unpredictability. Their healthy development, their happiness, and possibly their lives may depend on the awareness of nature that you awaken in them.

As a thoughtless economics and technology push our civilization ever faster toward some tumultuous and chaotic conclusion, teaching children to swim, in the larger sense of the instruction, takes on a special importance. What is required of us is not the impossible. We do not have to be perfect and we do not have to be prophets; we do not have to anticipate every new environment that our children may face. But as part of being attuned to nature, children should gradually be helped to understand that the world changes both spatially, which they know or guess, and with *time*, which few children comprehend. Temporal change is more rapid now than it has ever been in recorded history, and preparing children to recognize and cope with the frequently destructive changes in their environment is the swimming lesson at the start of this new century and the days ahead.

The children of wise parents will come to understand that massive and profligate use of resources, dysfunctional technologies, heedless abuse of nature, and inhumane, violence-ridden social structures promote a highly unstable civilization that cannot last many more years in its present form. They will not be paralyzed by fear because they will be ready for the change. They will be accustomed to reading books rather than watching television, sorting truth from lies, playing sports rather than electronic games, conserving rather than wasting, hiking rather than riding ATVs, practicing self-sufficiency rather than helplessly consuming, building communities rather than competing alone.

Easy to say, hard to do. Few parents, perhaps none, will completely meet this challenge. But there is no necessity to produce Olympic swimmers, just children who are comfortable and safe in the water. Although no Dr. Spock will lead us step by step along this parental path, maybe we don't need anything more than a willingness to experiment, a few general guidelines like the ones in the Talmud and other sources of tested wisdom, and the ability to accept whatever consequences may flow from our best efforts. And for the lucky parents, their children will help, making the journey together a delight, whatever the ultimate destination.

There is no magic formula for restoring our communities, no quick and painless silver bullet to cure the ills of this deeply damaged society. There is only hard and painstaking work with no promise of personal reward at some expected end. But the work itself, if done with our best skill and spirit, is a reward for our communities and ourselves, a reward that satisfies our deepest needs, a reward that never stops while the world lasts.

BIBLIOGRAPHY AND

SUGGESTED READINGS

The greatest part of a writer's time is spent in reading, in order to write; a man will turn over half a library to make one book.

Samuel Johnson,
quoted by James Boswell in
The Life of Samuel Johnson

Anyone who writes a book as wide-ranging as this one owes an enormous debt to the labors and ideas of many people. Here I acknowledge the contributions of others to my synthesis — and the pleasure that I have had in reading their works, some of which are listed below. (I have also cited my own prior books when I think they are relevant.) Annotation has been added where the title and the reference in the text itself do not explain why the book was included.

The Lies We Live:

Saul, John Ralston. *Voltaire's Bastards: The Dictatorship of Reason in the West.* New York: Vintage Books, 1992. The story of the development of the Age of Reason and its unexpected consequences, from Machiavelli and Loyola to Kissinger and Giscard d'Estaing, by a businessman, financier, novelist, and historian all in one.

Brainstorming

Ehrenfeld, David. *The Arrogance of Humanism.* New York: Oxford University Press, 1978; rev. ed., 1981.
Midgley, Mary. "Why Smartness Is Not Enough," in *Rethinking the Curriculum*, Mary E. Clark and Sandra Wawrytko, eds. 39–52. Westport, Conn.: Greenwood Press, 1990. Those who wish to explore further Midgley's profound understanding of the powers and limitations of modern science and the human mind will enjoy her books, including

Science as Salvation: A Modern Myth and Its Meaning (London: Routledge, 1992); and *Evolution as a Religion: Strange Hopes and Stranger Fears* (London: Methuen, 1985).

Orr, David W. *Earth in Mind: On Education, Environment, and the Human Prospect.* Washington, D.C.: Island Press, 1994. Much of the book is relevant, and especially the chapter entitled "Some thoughts on intelligence."

Pretending

Eco, Umberto. *Travels in Hyperreality.* San Diego, Calif.: Harcourt Brace Jovanovich, 1986. See the title essay in this collection, which is closely tied to the theme of pretending in America.

Ho, Mae-Wan. *Genetic Engineering: Dream or Nightmare?* 2d ed. Dublin: Gateway, 1999. This book and the following book by Lewontin, both written by distinguished scientists, provide the necessary careful evaluation of a field that receives a torrent of uncritical and often unwarranted adulation.

Jaenisch, Rudolf and Ian Wilmut. "Don't Clone Humans!" *Science* 291, p. 2552, 2001.

Lewontin, Richard. *It Ain't Necessarily So: The Dream of the Human Genome and Other Illusions.* New York: New York Review of Books, 2000.

Pliny's Natural History. trans. Philemon Holland, ed. Paul Turner. London: Centaur Press, 1962. When Pliny correctly described ostriches sticking their heads in the shrubbery and interpreted this as an ineffectual way of hiding, he did not know that only the male ostrich does it, and that he is not hiding but carefully turning the eggs he is incubating for a group of females in a communal nest on the ground. It is just as well that Pliny didn't understand the real reason for the ostrich's behavior, or we would have lost one of the best metaphors in the language.

Thornton, Joe. *Pandora's Poison: Chlorine, Health, and a New Environmental Strategy.* Cambridge: MIT Press, 2000.

The Magic of the Internet

Huber, Peter W., and Mark P. Mills. "Dig More Coal—The PCs Are Coming," *Forbes* 163(11), May 31, 1999, 70–72.

De Palma, Paul. "http://www.when_is_enough_enough?.com," in: *The Best American Science and Nature Writing 2000,* ed. David Quammen, 56–69. Boston: Houghton Mifflin, 2000.

Postman, Neil. *Building a Bridge to the 18th Century: How the Past Can Improve Our Future.* New York: Vintage Books, 1999. Chapter 3, "Technology," is the most relevant here, but reading only one chapter of Postman is like eating only one piece of popcorn.

Rochlin, Gene I. *Trapped in the Net: The Unanticipated Consequences of Computerization.* Princeton, N. J.: Princeton University Press, 1997.

Roszak, Theodore. *The Cult of Information: The Folklore of Computers and the True Art of Thinking.* New York: Pantheon, 1986.

Stoll, Clifford. *Silicon Snake Oil: Second Thoughts on the Information Highway.* New York: Doubleday, 1995. An Internet pioneer thinks hard about the best and worst uses of information technology.

———. *High-Tech Heretic: Reflections of a Computer Contrarian.* New York: Random House, Anchor Books, 2000.

Nothing Simple

Hardin, Garrett. "Nobody Ever Dies of Overpopulation," in *Stalking the Wild Taboo,* 2d ed. Los Altos, Calif.: William Kaufmann, 1978.

Hucker, Charles O. *China's Imperial Past: An Introduction to Chinese History and Culture.* Stanford, Calif.: Stanford University Press, 1975.

Miyazaki, I. *China's Examination Hell: The Civil Service Examinations of Imperial China.* trans. Conrad Schirokauer. New York: Weatherhill, 1976.

Sherlock, Nero, and Us

I have chosen the following mystery novels because I enjoyed them and because they are representative of the best work of the authors cited in this chapter.

Allingham, Marjory. *Police at the Funeral.* Harmondsworth, U.K.: Penguin, 1984.

Barron, Stephanie. *Jane and the Unpleasantness at Scargrave Manor.* New York: Bantam Books, 1996.

Christie, Agatha. *The Mysterious Affair at Styles.* New York: Bantam, 1961.

Freeling, Nicolas. *Sand Castles*. New York: Warner Books, 1989.

Freeman, R. Austin. *Mr. Pottermack's Oversight*. Mineola, N.Y.: Dover, 1985.

Gash, Jonathan. *The Sleepers of Erin*. Harmondsworth, U.K.: Penguin, 1984.

Hillerman, Tony. *A Thief of Time*. New York: Harper & Row, 1990.

James, P. D. *An Unsuitable Job for a Woman*. New York: Popular Library, 1972.

Keating, H. R. F. *Bats Fly Up for Inspector Ghote*. Chicago: Academy, 1974.

Melville, James. *A Sort of Samurai*. New York: St. Martin's Press, 1981.

Peters, Ellis. *One Corpse Too Many*. New York: Fawcett Crest, 1979.

Sayers, Dorothy L. *The Nine Tailors*. New York: Harcourt, Brace & World, [1934] 1962.

Stout, Rex. *In the Best Families*. New York: Bantam Books, 1962.

Van Gulik, Robert. *Necklace and Calabash*. New York: Charles Scribner's Sons, 1967.

Wrecking Our Society: A Manual

Williams, Terry Tempest. *Leap*. New York: Pantheon Books, 2000. An autobiographical journey through a painting by Hieronymus Bosch, in which the author, learning from Bosch's fifteenth century insights into people and nature, begins to understand the pathology in our modern existence.

Rejecting Gifts

Ehrenfeld, David. *Beginning Again: People and Nature in the New Millennium*. New York: Oxford University Press, 1995 (orig. 1993). See especially the chapter entitled "Forgetting."

Logsdon, Gene. "The Importance of Traditional Farming Practices for a Sustainable Modern Agriculture," in *Meeting the Expectations of the Land: Essays in Sustainable Agriculture and Stewardship*, ed. Wes Jackson, Wendell Berry, and Bruce Coleman. 3–18. San Francisco: North Point Press, 1984.

Williams, William Carlos. *Paterson*. Rev. ed. by Christopher MacGowan. New York: New Directions, 1995.

Adaptation

Carr, Archie. *Ulendo: Travels of a Naturalist in and out of Africa.* New York: Knopf, 1964.

Orwell, George. *The Road To Wigan Pier.* New York: Harcourt Brace Jovanovich, [1937] 1972.

Forecast: Chilly, Overcast, Light Drizzle, No People Left

Čapek, Karel. *R. U. R.: (Rossum's Universal Robots): A Fantastic Melodrama,* Paul Selver, trans. Garden City, N.Y.: Doubleday, Page & Co., 1923. In this play, Čapek, who coined the word "robot," lamented the folly of replacing humans with machines. The ending is happier than what we are likely to experience.

James, P. D. *The Children of Men.* Toronto: Penguin Books, 1992. Another take on the same theme: the preciousness of the human spirit.

Pseudocommunities

Austen, Jane. *Emma.* [1815]. Also see the citations for the chapter "Jane Austen and the World of the Community."

Gaskell, Elizabeth. *North and South.* London: Penguin Books, [1854–55] 1970.

———. *Wives and Daughters.* London: Penguin Books, [1864–66] 1986. Elizabeth Gaskell, like Jane Austen who preceded her, had an exceptional ability to describe the society and environment around her. Both of these books record changes in traditional community life, the first because of the new power of industrial development, and the second because of the growing importance of science.

Harlow, H. F., and R. R. Zimmerman. "Affectational Responses in the Infant Monkey," *Science* 130 (1959), 421–32.

McKibben, Bill. *The Age of Missing Information.* New York: Plume, 1993.

Mumford, Lewis. *The City in History: Its Origins, Its Transformations, and Its Prospects.* New York: Harcourt, Brace & World, 1961. See especially pages 511–13 ("Families in Space") for an explanation of why modern America has proved such fertile soil for pseudocommunities.

Trow, George W. S. *Within the Context of No Context.* New York: Atlantic Monthly Press, 1981; rev. ed., 1997.

Wilshire, Bruce. *The Moral Collapse of the University: Professionalism, Purity, and Alienation.* Albany: State University of New York Press, 1990.

Obsolescence

Gould, Stephen Jay. *Wonderful Life: The Burgess Shale and the Nature of History.* New York: Norton, 1989. One of the best, most thoroughly readable, and fascinating books on evolutionary history. Gould understands the danger of making sweeping evolutionary generalizations involving such concepts as obsolescence.

Lasch, Christopher. *The True and Only Heaven: Progress and Its Critics.* New York: W. W. Norton, 1991.

Sale, Kirkpatrick. *Rebels against the Future: The Luddites and Their War on the Industrial Revolution: Lessons for the Computer Age.* Reading, Mass.: Addison-Wesley, 1995.

Sturt, George [George Bourne, pseud.]. *Change in the Village.* New York: Augustus M. Kelley, [1912] 1969.

Sturt, George. *The Wheelwright's Shop.* Cambridge: Cambridge University Press, [1923] 1993. George Sturt, who also wrote under the name of George Bourne, wrote some of the best early descriptions of the impact of technological change—and the "meaningless" work that came with it—on the lives of people and their communities.

Tainter, Joseph A. *The Collapse of Complex Societies.* Cambridge: Cambridge University Press, 1988.

Social Evolution versus Sudden Change

Adams, Scott. *The Dilbert Principle: A Cubicle's-Eye View of Bosses, Meetings, Management Fads and Other Workplace Afflictions.* New York: HarperCollins, 1996.

Parkinson, C. Northcote. *Parkinson's Law: And Other Studies in Administration.* Boston: Houghton Mifflin, 1957. Still the best commentary on the curious, amusing, and frequently self-destructive structure and (mal)function of modern organizations.

Writing

Birkerts, Sven. *The Gutenberg Elegies: The Fate of Reading in an Electronic Age.* New York: Fawcett Columbine, 1994. Modern problems with writing are a direct result of the demise of the reading of books among a population hooked on TV and the Internet.

Mitchell, Richard. *The Graves of Academe.* Boston: Little, Brown, 1981. Mitchell is one of our most insightful—and ruthless—critics of bad writing.

Wodehouse, P. G. *The Most of P.G. Wodehouse.* New York: Simon & Schuster, 1960.

———. *Right Ho, Jeeves.* Harmondsworth, UK: Penguin, 1953. There is no point in criticizing bad writing without providing examples of how writing ought to be done. Wodehouse produced 100 books during his long life, all elegantly written and eminently readable. *Right Ho, Jeeves* is one of his best and funniest; the other volume listed includes some of his finest short stories.

Deadly Economics

Berry, Wendell. *Life Is a Miracle: An Essay against Modern Superstition.* Washington, D.C.: Counterpoint, 2000. Although couched as an extended book review, this stands alone as a tribute to life and a defense of the world view that credits science with its actual achievements, not its claims.

Midgley, Mary. *Evolution As a Religion: Strange Hopes and Stranger Fears.* London: Methuen, 1985.

Affluence and Austerity

Dobkowski, Michael N., and Isidor Wallimann, eds. *The Coming Age of Scarcity: Preventing Mass Death and Genocide in the Twenty-first Century.* Syracuse, N.Y.: Syracuse University Press, 1998.

Wachtel, Paul L. *The Poverty of Affluence: A Psychological Portrait of the American Way of Life.* Philadelphia: New Society, 1989.

Durable Goods

Kraft, Herbert C. *The Lenape: Archaeology, History, and Ethnography.* Newark: New Jersey Historical Society, 1986.

Quinn, John R. *Fields of Sun and Grass: An Artist's Journal of the New Jersey Meadowlands.* New Brunswick, N.J.: Rutgers University Press, 1997. Quinn has woven art, natural history, and human history together into a beautifully illustrated description of the meadowlands and its inhabitants, over time.

Spending Our Capital

Alexanyan, S. M., and V. I. Krivchenko. "Vavilov Institute Scientists Heroically Preserve World Plant Genetic Resources Collections during World War II Siege of Leningrad," in *Seed Savers 1992 Summer Edition*, Decorah, Iowa. See "Seed Savers Exchange" in citations for the chapter "I Reinvent Agriculture."

Bartlett, Albert A. "Forgotten Fundamentals of the Energy Crisis," *American Journal of Physics* 46, pp. 876–88, 1978. One of the most important scientific papers of the past half century, it is understandable by any well-educated person, including those who, like me, do not have special training in physics.

———. "Reflections on Sustainability, Population Growth, and the Environment—Revisited," *Renewable Resources Journal* 15 (Winter 1997–98), 6–23.

Campbell, C. J., ed. *The Coming Oil Crisis.* Brentwood, U.K.: Multi-Science Publishing Co. & Petroconsultants S.A., 1997. This 210-page paperback, based on a report originally prepared for the oil industry, brings together the thoughts of some of the world's foremost petroleum geologists. Historical in approach, it makes fascinating reading for the layman.

Duncan, Richard C., "The Peak of World Oil Production and the Road to the Olduvai Gorge," presented at the Pardee Keynote Symposium, Geological Society of America, Summit 2000, Reno, Nevada, Nov. 13, 2000. If you want to understand the underlying cause of the increasing frequency of electric blackouts, start here. Available from the

Institute on Energy and Man, 5307 Ravenna Place NE, #1, Seattle, Wash. 98105.

Youngquist, Walter. *GeoDestinies: The Inevitable Control of Earth Resources over Nations and Individuals*. Portland, Ore.: National Book Co., 1997.

Saving by Selling

Balick, Michael J., and Robert Mendelsohn. "Assessing the Economic Value of Traditional Medicines from Tropical Rain Forests," *Conservation Biology* 6 (1992), 128–30.

Geist, Valerius. "How Markets in Wildlife Meat and Parts, and the Sale of Hunting Privileges, Jeopardize Wildlife Conservation," *Conservation Biology* 2 (1988), 15–26.

Grigg, Gordon. "Kangaroo Harvesting and the Conservation of Arid and Semi-Arid Rangelands," *Conservation Biology* 3 (1989), 194–97.

Meylan, Anne B., and David Ehrenfeld. "Conservation of marine turtles," in *Turtle Conservation*, ed. Michael W. Klemens, 96–125. Washington, DC: Smithsonian Institution Press, 2000.

Vasquez, Rodolfo, and Alwyn H. Gentry. "Use and Misuse of Forest-Harvested Fruits in the Iquitos Area," *Conservation Biology* 3 (1989), 350–61.

Hot Spots and the Globalization of Conservation

Berry, Wendell. *The Unsettling of America: Culture and Agriculture*. San Francisco: Sierra Club Books, 1977. A classic account of what has gone wrong in the three-way relationship among modern society, the individual, and the earth.

Daly, Herman E., and John B. Cobb, Jr. *For the Common Good: Redirecting the Economy Toward Community, the Environment, and a Sustainable Future*. Boston: Beacon Press, 1989.

Dobson, A. P., J. P. Rodriguez, W. M. Roberts, and D. S. Wilcove. "Geographic Distribution of Endangered Species in the United States," *Science* 275 (Jan. 1997), 550–53.

Ehrenfeld, David, Reed F. Noss, and Gary K. Meffe. "Endangered Species 'Hot Spots,'" *Science* 276 (1997), 515–16.

Jackson, Wes. *Becoming Native to This Place.* Lexington: The University Press of Kentucky, 1994.

Kingsolver, Barbara. *The Poisonwood Bible.* New York: HarperCollins, 1998. A powerful historical novel about the triumph of dysfunctional contemporary culture over the once productive and durable community of the Congo.

Lake, Frank K. "In the Sacred High Country," in *Intricate Homeland: Collected Writings from the Klamath-Siskiyou,* ed. Susan Cross, 86. Ashland, Ore.: Headwaters Press, 2000.

Mander, Jerry, and Edward Goldsmith, eds. *The Case against the Global Economy: And for a Turn toward the Local.* San Francisco: Sierra Club Books, 1996. A fine collection of essays by the world's leading opponents of globalization.

Myers, Norman, Russell A. Mittermeier, Cristina G. Mittermeier, Gustavo A. B. da Fonseca, and Jennifer Kent. "Biodiversity Hotspots for Conservation Priorities," *Nature* 403, (Feb. 2000), 853–58.

Pimm, Stuart L., and Peter Raven. "Extinction by Numbers," *Nature* 403 (Feb. 2000), 843–45.

Smith, Page. *Killing the Spirit: Higher Education in America.* New York: Viking Penguin, 1990.

Soper, J. Dewey. "Solitudes of the Arctic," *Canadian Geographic Journal* 7 (1933), 103–15.

———. "The Lake Harbour Region, Baffin Island," *Geographical Review* 26 (1936), 426–38.

The Gingko and the Stump

Costanza, Robert, et al. "The Value of the World's Ecosystem Services and Natural Capital," *Nature* 387 (1997), 253–60.

Stevens, William K. "How Much Is Nature Worth? For You, $33 Trillion," *New York Times,* May 20, 1997, C1, C5.

The Death Penalty

Borkin, Joseph. *The Crime and Punishment of I. G. Farben.* New York: Free Press, 1978. Unbelievable if it weren't true; by a senior Nuremberg war trials prosecutor.

Bradsher, Keith. "Documents Portray Tire Debacle As a Story of Lost Opportunities," *The New York Times,* Sept. 11, 2000, A1, A23. Also see "Questions Raised about Ford Explorer's Margin of Safety," *The New York Times,* Sept. 16, 2000, C1, C4.

"Corporate Rights vs. Human Need," *Rachel's Environment & Health Weekly,* No. 677, Nov. 18, 1999. *Rachel's* newsletter, a major source of reliable, timely environmental and health news, is published by the Environmental Research Foundation, P.O. Box 5036, Annapolis, MD 21403-7036; erf@rachel.org.

Grossman, Richard. "Can Corporations Be Accountable?" (in 2 parts), *Rachel's Environment & Health Weekly,* nos. 609, 610, July 30, Aug. 6, 1998.

Marsh, George P. *Man and Nature: or Physical Geography As Modified by Human Action.* New York: Charles Scribner, 1864.

———. *The Earth as Modified by Human Action: A New Edition of Man and Nature.* New York, Arno Press, [1874] 1970. Epigraph from 1864 edition, p. 55; text quote from 1874 edition, pp. 53–54. The first modern, ecologically oriented book on conservation, written by a remarkable linguist, historian, lawyer, architect, diplomat, and geographer who predicted many of the most important environmental issues of the twentieth and twenty-first centuries.

Tanikawa, Miki. "Mitsubishi Admits to Broad Cover-up of Auto Defects," *The New York Times,* Aug. 23, 2000, A1, C19.

Relating to Nature in a Manmade World

Berry, Thomas. *The Dream of the Earth.* San Francisco: Sierra Club Books, 1988. A brilliant book by a historian of cultures who is both an advocate of sustainable communities and America's dean of religiously inspired environmentalists.

The Vine Cleaners

Leopold, Aldo. *A Sand County Almanac and Sketches Here and There.* New York: Oxford University Press, 1949. The twentieth century's most important book on people and conservation.

Meine, Curt, and Richard L. Knight, eds. *The Essential Aldo Leopold: Quo-tations and Commentaries.* Madison: University of Wisconsin Press, 1999. In this book, even more than in *A Sand Country Almanac,* we can see the evolution of Leopold's wisdom about the human need for nature.

Schumacher, E. F. *Small Is Beautiful: Economics As If People Mattered.* New York: Harper & Row, [1973] 1989. Although nature and the environ-ment are scarcely mentioned in this book, Schumacher well under-stood the mutually supportive relationship between "good work" and the natural world.

A Connoisseur of Nature

Begiebing, Robert J., and Owen Grumbling. *The Literature of Nature: The British and American Traditions.* Medford, N.J.: Plexus Publishing, 1990.

Finch, Robert, and John Elder, eds. *The Norton Book of Nature Writing.* New York: W. W. Norton, 1990. The two books cited for this chapter will provide a wealth of visions of nature quite different from those of my friend Claude.

Death of a Plastic Palm

Orwell, George. *The Road to Wigan Pier.* New York: Harcourt Brace Jo-vanovich, [1937] 1972.

———. *Coming Up for Air.* New York: Harcourt Brace Jovanovich, [1939] 1969.

———. "Some Thoughts on the Common Toad," in *The Collected Essays, Journalism and Letters of George Orwell,* vol. 4. New York: Harcourt Brace Jovanovich, [1946] 1968. Orwell was not only the most perceptive so-cial critic of his time, but also a competent naturalist. His comments on the substitution of plastic and steel for nature have not been sur-passed.

The following two books, mentioned in the chapter, will not help you with diseases of plastic trees; however, if you have real ones to care for they may be of use.

Harris, Richard. *Arboriculture: Care of Trees, Shrubs, and Vines in the Land-scape.* Englewood Cliffs, N.J.: Prentice-Hall, 1983.

Hartman, John R., Thomas P. Pirone, and Mary Ann Sall. *Pirone's Tree*

Maintenance, 7th ed. New York: Oxford University Press, 2000. [A more recent edition of the book cited in the text.]

Scientific Discoveries and Nature's Mysteries

Baker, Robin, chief contributing ed. *The Mystery of Migration.* New York: Viking, 1981. An interesting, beautifully illustrated book for the layman, with a historical survey and comprehensive coverage (from plants and invertebrates to humans) that more than compensate for its lack of information after 1981.

Bates, Henry Walter. *The Naturalist on the River Amazons: A Record of Adventures, Habits of Animals, Sketches of Brazilian and Indian Life, and Aspects of Nature under the Equator, during Eleven Years of Travel.* New York: Viking Penguin, [1863] 1989.

Berthold, Peter. *Bird Migration: A General Survey.* New York: Oxford University Press, 1993. An authoritative and readable account, with a good summary of Perdeck's experiment with starlings, pp. 140–42, and Papi's work with odor cues, pp. 163–65.

Berthold, P., A. J. Helbig, G. Mohr, and U. Querner. "Rapid Microevolution of Migratory Behaviour in a Wild Bird Species," *Nature* 360 (1992), 668–70. A commentary on this paper by William J. Sutherland, "Genes Map the Migratory Route," appears in the same issue of *Nature,* pp. 625–26.

Groot, C., "On the Orientation of Young Sockeye Salmon (*Oncorhynchus nerka*) during Their Seaward Migration out of Lakes," *Behaviour,* Supplement XIV (1965).

I Reinvent Agriculture

Fowler, Cary, and Pat Mooney. *Shattering: Food, Politics, and the Loss of Genetic Diversity.* Tucson: University of Arizona Press, 1990.

Jabs, Carolyn. *The Heirloom Gardener.* San Francisco: Sierra Club Books, 1984.

Nabhan, Gary Paul. *Enduring Seeds: Native American Agriculture and Wild Plant Conservation.* San Francisco: North Point Press, 1989.

Seed Savers Exchange. For up-to-date, interesting, and practical information about vegetable and fruit gardening, as well as access to a staggering number of "heirloom," nonhybrid seeds, you can't do better than join the Seed Savers Exchange (3076 North Winn Road, Decorah, Iowa 52101; 319 382 5990; seedsavers.org). There is a modest annual fee, for which you will be richly rewarded.

Wyndham, John. *The Day of the Triffids*. Harmondsworth, U.K.: Penguin, 1954.

Thinking about Breeds and Species

Bright, Chris. *Life out of Bounds: Bioinvasion in a Borderless World*. Worldwatch Environmental Alert Series. New York: W. W. Norton, 1998.

Cronk, Quentin C. B., and Janice L. Fuller. *Plant Invaders: The Threat to Natural Ecosystems*. London: Chapman & Hall, 1995.

Elton, Charles S., *The Ecology of Invasions by Animals and Plants*. London: Methuen, 1958. This is the first great book on the subject, and perhaps still the best. It is written by one of the finest ecologists of the twentieth century. The two concluding chapters on conservation are classics.

Meinesz, Alexandre. *Killer Algae: The True Tale of a Biological Invasion*. Chicago: University of Chicago Press, 1999. The biology and politics of a devastating species invasion. It reads like a mystery story, but it is definitely not fiction.

Mayr, Ernst W., *Evolution and the Diversity of Life: Selected Essays*. Cambridge: Harvard University Press, [1976] 1997. Everything you might want to know about what a species is. In places it can be difficult for laymen.

Tinbergen, Niko. *The Herring Gull's World*. New York: Basic Books, 1960.

Teaching Field Ecology

Brown, Tom, Jr. *Tom Brown's Field Guide to the Forgotten Wilderness*. New York: Berkeley, 1987. The "wilderness" is in and around your city or suburb, and Tom Brown is America's premier teacher of wilderness awareness and tracking. A fascinating guide to the daily enjoyment of nature—anywhere.

Comstock, Anna Botsford. *Handbook of Nature Study*. Ithaca, N.Y.: Cornell University Press, [1939] 1986. This book contains 887 pages of priceless, clearly written information about plants, animals, minerals, soil, weather, and the skies above.

Garber, Steven D. *The Urban Naturalist.* New York: John Wiley, 1987.

Kieran, John. *A Natural History of New York City.* Boston: Houghton Mifflin, 1959.

Leopold, Aldo. *A Sand County Almanac and Sketches Here and There.* New York: Oxford University Press, 1949.

Watts, May Theilgaard. *Reading the Landscape of America.* New York: Collier Macmillan, 1975. Watts describes the combination of detective ability and ecological knowledge required to understand the past and predict the future of the landscapes around us. See especially chapter 7.

More Field Ecology: Rightofway Island

Baum, L. Frank. *The Wonderful Wizard of Oz.* New York: Penguin Putnam [1900] 1984.

Niering, William A., and Richard H. Goodwin. "Creation of Relatively Stable Shrublands with Herbicides: Arresting 'Succession' on Rights-of-Way and Pastureland," *Ecology* 55 (1974), 784–95.

Kunstler, James Howard. *The Geography of Nowhere: The Rise and Decline of America's Man-made Landscape.* New York: Simon & Schuster, 1993.

A Walk in the Woods

Cronon, William. *Changes in the Land: Indians, Colonists, and the Ecology of New England.* New York: Hill and Wang, 1983.

Hoskins, W. G. *The Making of the English Landscape.* Harmondsworth, U.K.: Penguin Books, 1992.

Perlin, John. *A Forest Journey: The Role of Wood in the Development of Civilization.* Cambridge: Harvard University Press, 1991.

Rackham, Oliver. *The History of the Countryside.* London: Phoenix (Orion Books Ltd.) [1986] 1997.

Sauer, Leslie Jones, and Andropogon Associates. *The Once and Future Forest: A Guide to Forest Restoration Strategies.* Washington, D.C.: Island Press, 1998. An exciting book that describes what the temperate forests of the eastern United States once were, and what they might, with wise restoration, become again, not only in rural areas but in the suburbs and cities as well.

Wessels, Tom. *Reading the Forested Landscape: A Natural History of New England*. Woodstock, Vt.: Countryman Press, 1997.

Van Creveld, Martin. *The Transformation of War*. New York: Free Press, 1991. A distinguished military historian looks ahead at the nature of the likely military conflicts of the twenty-first century.

Degrees of Intimacy

Peacock, Doug. *Grizzly Years: In Search of the American Wilderness*. New York: Henry Holt, 1995.

Sauer, Peter, ed. *Finding Home: Writing on Nature and Culture from* Orion *Magazine*. Boston: Beacon Press, 1992. The essays in this book explore many of the exciting ways that people can develop an intimacy with nature.

Teale, Edwin Way, ed. *The Wilderness World of John Muir*. Boston: Houghton Mifflin, 1954. Contains some of the best of Muir's writing, including many extracts that portray his unself-consciously fearless intimacy with nature.

Restoring the Community

Berry, Wendell. *Life is a Miracle; An Essay against Modern Superstition*. Washington, D.C.: Counterpoint, 2000.

The Utopia Fallacy

Berlin, Isaiah. "The Pursuit of the Ideal," in *The Proper Study of Mankind: An Anthology of Essays*, Henry Hardy and Roger Hausheer, eds., 1–16. New York: Farrar, Straus and Giroux, 1998.

Curry, Patrick. *Defending Middle-Earth: Tolkien, Myth and Modernity*. New York: St. Martin's Press, 1997. Tolkien's brand of non-utopian hope, which I share, is beautifully and movingly explained by this British scholar. Note especially the last chapter: "Conclusion: Hope Without Guarantees."

Diamond, Jared. *Guns, Germs, and Steel: The Fates of Human Societies.* New York: Norton, 1997.

Hannum, Hildegarde, ed. *People, Land, and Community: Collected E. F. Schumacher Society Lectures.* New Haven: Yale University Press, 1997. The essays in this collection contain many practical, non-Utopian ideas for community restoration, from local currencies to the Garden sProject.

Maimonides. *Pirkei Avot: With the Rambam's Commentary,* Rabbi Eliyahu Touger, ed. and trans. New York: Moznaim Publishing Corp., 1994. There are many editions of *Pirkei Avot (The Ethics of the Fathers).* This one, with Maimonides' (the Rambam) commentaries and other material by the medieval philosopher, is particularly interesting.

Mander, Jerry, and Edward Goldsmith, eds. *The Case against the Global Economy: And for a Turn toward the Local.* San Francisco: Sierra Club Books, 1996.

Saul, John Ralston. *Voltaire's Bastards: The Dictatorship of Reason in the West.* New York: Vintage, 1992.

Schumacher, E. F. *Small Is Beautiful: Economics As If People Mattered.* New York: Harper & Row, 1989.

Solzhenitsyn, Aleksandr I. "To Tame Savage Capitalism," *The New York Times,* Nov. 28, 1993, Op-Ed page.

Tolkien, J. R. R. *The Lord of the Rings.* 3 vols. London: George Allen & Unwin Ltd, 1954.

Van Creveld, Martin. *The Transformation of War.* New York: Free Press, 1991. A distinguished military historian looks ahead at the nature of the likely military conflicts of the twenty-first century.

Traditions

Bernstein, Ellen, ed. *Ecology and the Jewish Spirit: Where Nature and the Sacred Meet.* Woodstock, Vt.: Jewish Lights Publishing, 1998. This book and the two that follow, although not confined to the subject of traditions, contain much useful material on the interaction between religious tradition and ecology.

Carroll, John E., Paul Brockelman, and Mary Westfall, eds. *The Greening of Faith: God, the Environment, and the Good Life.* Hanover, N.H.: University Press of New England, 1997.

Dien, Mawil Izzi. *The Environmental Dimensions of Islam.* Cambridge, U.K.: Lutterworth Press, 2000.

Encyclopedia Judaica. Jerusalem: Keter Publishing House, 1972. A comprehensive and scholarly source of information on all things Jewish.

Hobsbawm, Eric, and Terrence Ranger, eds. *The Invention of Tradition.* Cambridge: Cambridge University Press, 1983.

Schaffer, Arthur. "The Agricultural and Ecological Symbolism of the Four Species of Sukkot," *Tradition* 20 (1982), 128–40.

Jane Austen and the World of the Community

Austen, Jane. *The Complete Novels of Jane Austen.* New York: Modern Library, n.d. There are many paperback editions of Jane Austen's six completed novels; unlike this collection, most have introductions and editor's notes.

Kunstler, James Howard. *Home from Nowhere: Remaking Our Everyday World for the Twenty-first Century.* New York: Simon & Schuster, 1996.

Le Faye, Deirdre. *Jane Austen's Letters,* 3d ed. New York: Oxford University Press, 1995. Any spelling irregularities in quotes from Austen's letters reflect her own orthographic preferences.

Tanner, Tony. *Jane Austen.* Cambridge: Harvard University Press, 1986.

Vitek, William, and Wes Jackson. *Rooted in the Land: Essays on Community and Place.* New Haven, Conn.: Yale University Press, 1996.

Universities and Their Communities

Kozol, Jonathan. *Amazing Grace: The Lives of Children and the Conscience of a Nation.* New York: HarperCollins, 1996.

Orr, David W. *Earth in Mind: On Education, Environment, and the Human Prospect.* Washington, D.C.: Island Press, 1994. A sweeping and remarkably informed vision of the path that education must take if the best parts of our culture and environment are to survive in the decades ahead.

Rowe, Stan. "Role of the University," in *Home Place: Essays on Ecology,* 129–38. Edmonton, Alberta: NeWest, 1990.

Smith, Page. *Killing the Spirit: Higher Education in America.* New York: Viking Penguin, 1990.

Wilshire, Bruce. *The Moral Collapse of the University: Professionalism, Purity, and Alienation.* Albany: State University of New York Press, 1990.

Yunus, Muhammad, with Alan Jolis. *Banker to the Poor: Micro-Lending and the Battle Against World Poverty.* New York: Public Affairs, 1999. Perhaps the most hopeful book on this list. Yunus shows that it is possible to change the world for the better on a grand scale.

An Invalid's Guide

Sharansky, Natan. *Fear No Evil.* trans. Stefani Hoffman. New York: Random House, 1988.

Swimming Lessons

Kozol, Jonathan. *Ordinary Resurrections: Children in the Years of Hope.* New York: Crown, 2000. About the power of children to be children even in a harsh, urban environment. Seen through the eyes of the children themselves, as told to the author. Shows not only children learning from adults, but adults learning from children.

Nabhan, Gary Paul, and Stephen A. Trimble. *The Geography of Childhood: Why Children Need Wild Places.* Boston: Beacon Press, 1994.

Postman, Neil. *Building a Bridge to the 18th Century: How the Past Can Improve Our Future.* New York: Vintage Books, 1999. The chapters on children and education are particularly relevant to my discussion. Postman's writing is lucid and engaging, and his arguments are profoundly important.

Spock, Benjamin, and Steven J. Parker. *Dr. Spock's Baby and Child Care,* 7th ed. New York: Pocket Books, 1998. A later and thicker edition than the one I used—it is good to see that this invaluable book is being revised and kept in print.

INDEX

low-intensity conflict, 175. *See also* wars, local

Luddites, 46

lulav. See palm frond

Lysenko, Trofim Denisovich, 77

McClintock, Barbara, 9

MacArthur Foundation, 94

MacDonald, Kristi, 96

Macedonia, 175

MacMillan Bloedel, 164

mad cow disease. *See* spongiform encephalopathy

Madison, James, 175

magnetic fields, navigation by, 134–135

magnetic resonance imaging (MRI), 173

Mahican Indians. *See* Indians, American

Maimonides, 179

Maine, 41, 67, 95

mammals, 50–51

Manhattan, 36–37, 41, 108, 120–121

Manitoba, 118

Mansfield Park, 182–183, 187

Mao Tse-tung, 21

maple, Norway. *See* trees

maple, red. *See* trees

maple, silver. *See* trees

market, world, 84, 86–87

Marsh, George P., 108, 111

Marshall, John, 111

Marxism, Marxist theory, 174

Massachusetts, 192

Massachusetts Institute of Technology, 12

materialism, 188

mayfly, 149

Meares Island, 164

mechanization, 38

medicines, traditional, 86

melons, 33

Melville, James, 27

Mendelsohn, Robert, 86

mensch, 120

Meta Incognita Plateau, 89, 92

Mexico, 193

mice, field, 155

Middle East, 208

Midgley, Mary, 6, 65

midrash, 178, 181

migration, bird, 131–134

Milky Way, 134

millet, 77

Millstone River, 101, 152

Mitsubishi Motors, 112

Mittermeier, Russell, 93–94

Mongolia, 95

Monsanto, 196

moorlands, 124

Moriori, 174

Moscow, 43

Moses, Law of, 175

mosquitoes, Asian tiger, 144

mosses, 90, 149

moths, gypsy, 146

Muir, John, 165

multinationals. *See* corporations

Mumford, Lewis, 45, 148

mussels, zebra, 146

mustard, red, 138

Myers, Norman, 92–96

myrtle, 177–180

Nanticoke Indians. *See* Indians, American

National Aeronautic and Space Administration (NASA), 173

nationalism, 181